God and the New Metaphysics

"This book is a vigorous plea for a new metaphysics capable of revealing being and becoming as the texture of both the world and the nature of God."
Maurice Boutin, Ph.D., J.W. McConnell Professor of Philosophy of Religion, McGill University

"*God and the New Metaphysics* is a handy guide ('road map') for anyone venturing on philosophical excursions into the big questions surrounding the origins of the universe itself, and of life within it. It succinctly summarizes scientific advancements and their implications for the way we view the world, since (as Herb Gruning writes) 'All epistemological activity ... functions worldviewishly.' In addition to dealing with the development of the philosophy of science in the context of the clashes between science and religion, the author suggests interesting lines of inquiry to help resolve the questions which will remain in the reader's mind at the end of the book, as they do in philosophy and science to the present day."
Francine McCarthy, Ph.D., Professor of Earth Science and Great Books/ Liberal Studies, Brock University

"What can abide and what must change in our understanding of God and nature? Walking the ever-shifting boundary lines between religion and contemporary science and drawing on rich resources to be found on both sides, Gruning has written a well-informed, insightful, and adventuresome book, addressing this question. And while refusing to oversimplify, he manages to write lucidly about deep and complex matters."

John C. Robertson, Ph.D., Professor Emeritus, Department of Religious Studies, McMaster University

"In lucid and sparkling prose, Herb Gruning provides the inquisitive layperson with a solid summary of the history and present state of the natural sciences, together with a panoply of the most significant meta-scientific reflections on the meaning and import of those sciences for the higher purposes of human life. *God and the New Metaphysics* offers more than a road map through the highways and byways of science, philosophy, and theology. It is a traveler's guide through adjacent territories that have been explored by other travelers, with lively commentary on the usefulness of the older guidebooks. Above all, it identifies the limits of the previous explorations, and so reveals the openness of the horizons of knowledge, instilling in the reader a sense of further adventure at the edge of the known world."

James Lawler, Ph.D., Philosophy Department, SUNY at Buffalo

"The author gives an excellent account of the controversy surrounding science and religion as alternative avenues to truth and argues persuasively how science in general has failed to unravel some of the deep mysteries of the universe such as the origin of the universe and of life, among others. The book makes interesting reading with the author's critical look at traditional theology with respect to its stand on the nature of god and purpose of creation, and his suggestion for a new direction in this regard along the lines developed by Alfred North Whitehead and others who consider the universe in all its diversity as a Divine Process—a view cosistent with that of modern physics where the cosmic reality is looked upon as a play of cosmic energy. While it is a valuable addition to one's collection, the book provides useful material for a course in science and religion."

Gowdar Veeranna, economist, Winnipeg, Canada

God and the New Metaphysics

HERB GRUNING

BLUE DOLPHIN PUBLISHING

Copyright © 2005 Herb Gruning
All rights reserved.

Published by Blue Dolphin Publishing, Inc.
P.O. Box 8, Nevada City, CA 95959
Orders: 1-800-643-0765
Web: www.bluedolphinpublishing.com

ISBN: 1-57733-161-3

Library of Congress Cataloging-in-Publication Data

Gruning, Herb.
  God and the new metaphysics / Herb Gruning.
    p. cm.
  Includes bibliographical references and index.
  ISBN 1-57733-161-3 (pbk. : alk. paper)
  1. Cosmology. 2. God. 3 Metaphysics. I. Title.

BD511.G78 2005
211—dc22
                                    2005015927

Printed in the United States of America

10   9   8   7   6   5   4   3   2   1

Contents

Introduction — 1
- *Speculations on the Way Things Might Be* — 4
- *Can We Get There from Here?* — 5

Chapter 1. Religion and Science — 8
- *The Nature of Religion in Scientific Terms* — 12
- *The Nature of Science in Religious Terms* — 14
- *Enter the Philosophy of Science* — 17
- *The Recent Situation* — 18
- *A Diagnosis of the Enduring Problem* — 23
- *The World According to Gilkey* — 26
- *The Nature of Nature* — 29

Chapter 2. The New Physics Applied to the Old Nature — 32
- *The Quantum World* — 32
- *The Relativistic World* — 34
- *Einstein Versus Bohr* — 36
- *The Anthropic Principle* — 42
- *How It All Began: The Early Years* — 45
- *The Story Thus Far* — 47

Chapter 3. The Reenchantment of Nature — 53
- *Process as Reality* — 56
- *A Sporting Chance* — 65

Chapter 4. **How It All Unfolded: The Later Years**   73
    Does God Wear a Designer Hat?   82
    What's the Purpose?   90

Chapter 5. **Have We Made Any Progress?**   96
    Stephen Jay Gould   97
    James Lovelock   105
    Rupert Sheldrake   114
    Living in Resonance   121
    Robert O. Becker   130
    Pierre Teilhard de Chardin   132

Chapter 6. **New Vistas**   138
    The Past as Causal   138
    Whitehead and My Discomfort Zone   145
    David Bohm   149
    The Combination Approach: Syncretic or Synergetic?   156
    The Forces of Nature   160

Chapter 7. **Which Way to Turn?**   164
    When in Doubt, Venture Out   170
    On a Personal Note   172
    An Initial Final Word   175

Appendix 1. **God by Any Other Name**   176

Appendix 2. **Time for a Change**   180

Notes   188

Bibliography   190

Index   197

Acknowledgments

My thanks to those who read the manuscript proofs: Dr. Maurice Boutin, Mr. Prasad Gowdar, Dr. Norman King, Dr. Stanley Krippner, Dr. James Lawler, Dr. Francine McCarthy, Dr. John Robertson, and Mr. Gowdar Veeranna. Thanks also to Candida Hadley as well as the staff at Blue Dolphin for their editorial work.

This book is dedicated to my wife, Alice,
for her constant companionship and encouragement.

Introduction

HEAVEN KNOWS if our entrance into the year 2000 will be remembered in North America for these two things: the Y2K computer bug and the price of petroleum. Concerning the first, much anxiety was generated by the Y2K scare, with projections of catastrophe on a global scale both on the ground and in the air. The result, however, was much less than originally forecast and initially feared, hardly producing a ripple in overall technical service. Calm was restored and normal activities resumed largely as before. There were some computer systems, though, being insufficiently Y2K compatible, that did experience some glitches. One of those that did not escape unscathed was my own laptop, which dealt with the turn of the year by a curious form of adaptation—it inverted one of the digits in 1999. Unprepared for something called a year 2000, it elected to reckon with the new year by reverting to 1969. Some of those to whom we sent emails thereafter noticed the error and thought it was old mail and did not bother to open and read them, as though PCs and the emailing which the internet affords was already in full swing back when humans first strolled on the moon. If those correspondents of ours were to be asked, perhaps they would have counselled us that it is time to invest in a new computer.

Regarding the second item, the price of gas at the pumps reached new heights in the year 2000. This was not so much a reflection of the scarcity of the resource, what in the 1970s was referred to as a potential oil shortage, as the unwillingness, for whatever reason, of oil-producing nations to boost production.

2 God and the New Metaphysics

Petroleum is held to be a non-renewable resource, the reserves of which are believed to be steadily dwindling. The 1970s made us aware that the amount of oil in the planet is finite, alerting us that we also have a limited time to envision what a new source of power ought to look like. Yet the predicament is felt most acutely when it reaches into our pockets, prompting us to deal with a situation before it reaches crisis proportions.

A growing dissatisfaction with the cost of filling up might move us toward avidly seeking for alternate means of locomotion than the conventional internal combustion engine. Automobiles that run on propane have been devised, the manufacture of solar-powered vehicles is underway, albeit still in the embryonic stage, and hybrid engines powered by a blend of gas and a form of alcohol have appeared on the scene at least to smooth the transition from ordinary petroleum to something else. With the current price of gas, perhaps there is enough incentive to invest in a new way of getting around.

Using a little bit of imagination, one could also diagnose that a parallel situation obtains on the theological side of life. For centuries, views of God have been modified to comport with changing philosophical climates. The scientific and industrial revolutions, for instance, inspired a mechanical metaphor for the divine. Charles Darwin's legacy provoked a picture of the deity, if one could still be retained, as one that did not churn out species like an assembly-line worker, but in certain respects let nature take its course. Views of God were further modified to account for a divinity that would countenance two world wars, a holocaust and a series of mass suicides. Last century, portrayals of deity required alteration with the dawning of the awareness that humans were not as perfectible as previously supposed. This century, with the genome project and its applications and implications, human perfectibility has the prospect of potentially being manufactured. Perhaps biology can now correct what education could not. And in this setting, views of divinity will need to adapt again in order to survive.

Attempts have been made to provide new pictures of God once the former ones became outmoded. Much like in Thomas

Kuhn's description of scientific revolutions, a period of normal theology could be said to prevail until such time as anomalies or counterexamples build up, if Kuhn's analysis is correct, to induce the upsetting of the current theological paradigm and the establishment of a new one. The old portrayal of God evidently could not come to terms with more recent developments and hence required renovation.

Arguably, the most extensive attempt to foster a view of God that could address twentieth century ills was supplied by Alfred North Whitehead. It has been more than three-quarters of a century since his first major work in metaphysics was published, and since that time the world has once again covered new ground. The proponents of process thought, as it is named, affirm that the intervening time and the changes it has seen can still be incorporated into the Whiteheadian program. This may or may not be the case. In order to be thorough, we must at least ask if there are not also fragilities building up in the Whiteheadian framework. Whitehead would have been the first to admit that a system which is truly processive would itself experience process. This might extend to ideas as well as entities. A theoretical scheme that goes beyond Whitehead's original formulation would be applauded by him, since he cautioned us that once a metaphysical system has been sculpted it should then be mistrusted. To do so is to be in the Whiteheadian tradition. Are we entering such a phase?

Not only process thinkers but many others have grown increasingly disillusioned with classical conceptions of God. Whitehead himself experienced the loss of his son in the First World War, and the extent to which his framework incorporates a divinity is his effort to come to grips with this tragedy. Others might agree with the sentiment of reformulating their perspective on God but have traversed a different path than Whitehead. Whatever direction has been taken, there are many who have been driven to take one of a number of forks in the road. For them the forks will not go away and a decision must be reached as to which way to go. The older image of God has outlived its usefulness for many and a new way must be forged. For these sojourners, at least, perhaps it is time to invest in a new vision of God.

4   God and the New Metaphysics

Speculations on the Way Things Might Be

Far be it from me to announce that any one of us has unearthed the answers to the ultimate questions of the universe such that the nature of things is fully and finally settled. I repeat, far from it. Hence the modest claim for this section's heading.

At best, ruminations about the perennial problems in philosophy and religion amount to hopeful conjectures. A school of thought may provide what is taken as definitive, but it is not conclusive. The shifting winds of fashion and tradition will eventually disturb and overturn what was previously thought to be firmly established. The work before you is no different. What is worse, it does not even presume to approach the status of being definitive. It is merely comprised of reflections on what has impressed itself upon the present author's thinking at some point. No more, no less.

As one framework for thinking and living has presented itself for scrutiny, it has won favor in my eyes for a time. Yet after a while, the tide turns and it no longer holds the same import for me. The accumulation of weighty objections finally reached a critical mass, whereupon I discovered that I am no longer a card-carrying member of a certain philosophical or religious fraternity. This is not intended to imply that these schemes were either accepted or discarded wholesale at any point. Some aspects of each approach, in fact, continue to carry weight in my estimation.

Some features of a world and life view also remain steadfast for me. This may or may not be true in the experience of others. As for myself, I continue to have a God-consciousness, that is, I recognize that I dwell in a universe which exhibits God's fingerprints. I cannot even recall a time when I ever thought that reality was devoid of a divinity. My conviction was and still is that there is Someone in charge of the cosmos and who also has a hand in it as well as a claim on my life. Another person, however, may come to the same universe, perhaps with even similar questions, but will retreat from the notion that traces of divine management can be found there. This means that the two of us actually inhabit two different thought worlds. There will be a certain commonality between us—a region of intersection or overlap—but also much that is foreign. Some elements of one vision will resonate with those of another view, others will not. There will be contours of our commitments that converge as well as diverge. This is to be expected; no two people are alike, and neither are they totally different.

Each person is as unique as the worldview he or she subscribes to. Commitments, though, can change from time to time. The foundation of my own convictions appear to endure and remain intact and for this I am grateful. The structure or edifice built on top of the foundation, however, is under construction. I am currently undergoing a renovation of the upper floors.

For those individuals who consider themselves to be on route, that is, to be on a philosophical or theological journey as opposed to already having arrived at a resolution to their cosmic scale issues, this investigation might prove helpful if the reader comes to his or her universe with related questions. This volume examines metaphysical and cosmological proposals for an alternate vision of reality. It also amounts to the author's personal progress report.

The title of this work is a deliberate play on physicist Paul Davies' volume *God and the New Physics*. The difference is that whereas his approach is a scientific one, my study comes to the subject of God and cosmology from a philosophical and theological angle, while at the same time remaining sensitive to scientific concerns. Another variation is that the present volume is a type of travel log to highlight the idea that if one is intent on venturing out into the cosmos philosophically, then it is best to bring along a road map. Some of this ground, or should I say space, has already been mapped out, while other territory has yet to be uncovered, and still others require cartographic revision. My aim is to scan the topography.

Can We Get There from Here?

The late Douglas Adams, in the first instalment of his now famous "five-part trilogy" entitled *The Hitchhiker's Guide to the Galaxy*, offers some advice to those who would venture out as galactic travellers. He counsels those brave and adventurous souls to carry with them in their accouterments a towel. Along with one's other personal effects, towels are needful, he suggests, for a number of reasons, for which the reader is directed to the third chapter of the first volume of said work. To my thinking, the intrepid galactic sojourner is better equipped with a map. For Adams, the Guide itself, which is a sort of palm pilot version of the *Encyclopedia Galactica*, serves such a purpose. Yet presently lacking a device of this type, we are left with the standard approach.

Maps describe a terrain and travel logs recount the tale of a journey and recommend certain strategies as well as warn against others. This excursus cannot be considered either of the two since the terrain is often uncharted and the region untrodden, at least by *Homo sapiens sapiens*. What follows then amounts to where the present author and those he investigates have gone in their imaginations.

The themes treated in the subsequent pages derive largely from the following grid. According to conventional scientific wisdom, our current universe began with a Big Bang, which yielded the eras, in order of appearance, of physics, chemistry and biology. The origin of the universe remains a scientific mystery, but once it was under way, additional mysteries surfaced. There is a gap in our understanding, presently at least, between the world of chemistry and the onset of biology. How life arose from non-life has not been resolved. Then there is also a gap between biology and psychology, namely how an entire host of metaphysical categories arose, or at least the alleged realities to which they point. These include mind, soul, spirit, consciousness, self-consciousness, awareness, and self-awareness. Mysterious too is not only how the world began but how it will end.

A further concern is how many classes of things there are in the universe—one or two? If the former, there are three main types of monist approaches: materialistic, where there is matter only; idealistic, where there is just mind; and the panexperientialistic view of certain process thinkers in which there is experience only. If the latter, there is a dualistic outlook where both matter and mind are entertained, despite uncertainty as to how they can interact.

While the current analysis cannot hope to unravel these mysteries once and for all, it engages the thought of some of those who begin to disentangle us from the metaphysical knots in which we have tied ourselves. In sequential order we will consult the work of Whitehead, James Lovelock, Rupert Sheldrake, Pierre Teilhard de Chardin, David Bohm, and sample the thought of several others for the insights they can bestow. While any of these offerings may not hold out long-term promise, perhaps a combination of them might. Hence the uncharted nature of the terrain.

So let the journey commence with somewhat familiar territory.

CHAPTER ONE

Religion and Science

[Galileo speaking] But, gentlemen, [humans] can misinterpret not only the movements of the stars, but the Bible too. (Brecht 1960, 71)

Things have taken a strange turn in recent years; almost the full circle from Galileo's famous struggle with the theological establishment. It is the scientific establishment that now forbids heresy. (Lovelock 1979, vii)

GALILEO FOUND AN ADVERSARY in the Church but not for reasons normally stated. Popular accounts of religion and science controversies depict religious authorities at Galileo's time as fearful of theories in opposition to what they interpreted as scriptural teaching—the battle lines being drawn between what the Bible says and what science discloses. In actuality, the Church was not alarmist when it came to the Copernican views that Galileo was advocating so long as they were taught as hypotheses only. What was at issue was the allegiance to authority or rule; which of the two—science or the Church—commanded governing power? Both could not enjoy final authority. As Virginia Stem Owens declares, "[w]hat the Church denied was that a hypothesis, *any* hypothesis (which is, after all, only an intellectually conceived model), could be identical with ultimate truth" (Owens 1983, 27). The tension arising between Galileo and the Church resulted from mistaking models for reality, or what Whitehead refers to as "the fallacy of misplaced concreteness." This amounts to "mistaking the abstract for the concrete" which is what happens if, say, the cosmic map mentioned above were mistaken for the very terrain it signifies (Whitehead 1967b, 51).

Theorizing then has its potential hazards: "Because our representation of reality is so much easier to grasp than reality itself, we tend to confuse the two and to take our concepts and symbols for reality" (Capra 1983, 35). Difficulties ensue whenever "[a] technique of investigation [is] on its way to becoming a total account of the world," that is, as a method becomes translated into a metaphysic (Barbour 1966, 36-7). As Owens continues, "what was at stake was not a new theory of nature but a new nature of theory. When scientists began to take their models for the real thing, the Church brought the weight of her authority against their intellectual presumption" (Owens 1983, 27).[1]

The history of religion and science controversies has been sufficiently recounted elsewhere. Yet there is more than one way to write history. Every account of the development of and movements within civilizations bears a certain flavor. The one who attempts to convey "what happened" in a specific period under consideration will do so by framing it according to a certain motif. The very act of framing is a creative one which enables the reader to be guided by a thread or threads running through the account. These threads assist in making sense of the history being presented.

In such a creative undertaking, though, the materials scrutinized by the historian do not themselves come ready-made with these motifs or threads. One author sees the unfolding of history in one way, another author views it differently. Each one approaches similar sources, and similar historical questions might even be posed, but the issues raised do not immediately point to a definitive explanation. Different historians will couch the same event in a variety of ways depending on what they regard as the most important aspects and themes.

Themes are not so much unearthed as they are invented. Histories are not so much reported as they are drawn or crafted.

Threads or patterns do not emerge as much as they are imaginatively utilized as tools for descriptive purposes. And what this describes is a creative process. More on this momentarily.

A history of the interaction (assuming there is any) between organized, institutional religion and the rise of modern science permits different yet related canvasses upon which accounts may be painted. John William Draper (1811-82) and Andrew Dickson White (1832-1918) are two authors who offer a stormy rendition of the relation between religion and science, though along differing lines. From their dates it can be appreciated that the idea that religion and science manifestly conflict is a relatively new one—it has a short history (thus far) since it was popularized from about the last quarter of the nineteenth century.

Draper, a U.S. scientist and historian, wrote *A History of the Conflict Between Religion and Science*, which was first published in 1874. According to the late Stephen Jay Gould, himself quoting a recent work by the historian J.B. Russell,

> it was the first instance that an influential figure had explicitly declared that science and religion were at war, and it succeeded as few books ever do. It fixed in the educated mind the idea that "science" stood for freedom and progress against the superstition and repression of religion. Its viewpoint became conventional wisdom. (Gould 1999, 121)

As an interesting historical aside, it was also Draper who gave the address about the publication of Charles Darwin's theory of evolution that T. H. Huxley and Bishop Wilberforce responded to, their debate having become the stuff of legend (Gould 1995, 48).

White, a U.S. educator and the first president of Cornell University, authored *The Warfare of Science*, first published in 1876 and enlarged in 1896. Then by 1905 it appeared in two volumes under the title *A History of the Warfare of Science with Theology in Christendom*—the terms becoming increasingly inflammatory as time goes on. Both historians describe religion and science as antagonistic toward each other, but give an alternate impetus for it as well as a different vision for the future. White was a devout theist who hoped to save religion from its dogmatic turn and salvage the true faith. He was trying to build bridges between religion and science, although his writings were seen as antagonistic toward religion (Gould 1999, 101-2). Draper, on the other hand, was a physician who dabbled in history and definitely had an axe to grind with religion. He did, however, trust that science could be buttressed by religion but only with that pecking order firmly in place (Gould 1999, 103).

The problem, though, is not with religion but "particular embodiments" of it, more precisely "characterized as dogmatic theology," where religion becomes an impediment to progress in science. In the Galileo case, he and the Church were most assuredly at odds, but the point to be made here is that the notion that it is the nature of science and the character of the Church to inevitably conflict or be inherently inimical toward each other, whether accurate or not, is a recent view—traceable back to Draper and White as popularizers of this idea.

The intent of the present analysis is not to diagnose how religion and science or God and the world interact, if at all, but to highlight potential steps toward their unity. Whitehead is one author who claims that the two can be considered as having been woven from the same fabric. In his terms, "[s]cience suggest[s] a cosmology; and whatever suggests a cosmology, suggests a religion" (Whitehead 1974, 136). (At the time of his writing, a cosmology was taken as equivalent to a worldview.) Ways to a knowledge and understanding of God for Whitehead includes science.

The focus of the next section will begin with a statement of the extent to which theology can be regarded as a scientific discipline followed by an examination of any religious elements in scientific methodology. The issue to be resolved is whether both religion and science are on similar or analogous philosophical footing, precisely because *subjects* engage in both. Then we will determine how current scientific approaches display openness to transcendence or at least find themselves at home with metaphysical descriptions of nature. Previewing the remainder of this study, we will investigate the contributions that Whitehead and others make to the discussion and then consider tentative proposals for fruitful directions which the field of religion and science might take.

12 *God and the New Metaphysics*

The Nature of Religion in Scientific Terms

> *[B]iblical principles may play a legitimate role as framing principles, but ... they are not data, nor are they theories in the scientific sense.* (Wright 1989, 65)

Both religion and science garner data from experience. By reflection on experience, ideas on a grand or cosmic scale can be cultivated. Each avenue appears to offer a proposal for the quest for piecing together the human (and divine) puzzle. Both science and religion then have more in common than we have been led to expect from the accounts of their conflict.

We are assisted in our attempt to situate the introduction of the notion of God into philosophical and theological systems by the Spanish philosopher Xavier Zubiri. He locates in the ancient Greek thinker Aristotle the idea that "for the first time in the West, the inclusion of the theme of God in a system of speculation is realized" (Zubiri 1981, 304). Subsequent to Aristotle, the second instance for Zubiri occurs in the mystery religions where speculative knowledge gave way to the ecstatic as the "supreme form of intellection." This development "from rationalism to mysticism saw its culminat[ion] in Neoplatonism." In addition to the Greek and oriental conceptions of the divine came the Christian movement. At that point "a third type of problem about God is injected into Greece," namely aspects of the Judeo-Christian deity. This in turn inspired "the creation of Greek theology, the theology of the first Fathers of the Church; and above all, the creation of the first speculative and systematic theology, that of Origen" (Zubiri 1981, 304-5). Zubiri then discloses another divide which occurred one and a half millennia later. Since the above system of speculative reason was ultimately "incapable of elevating itself from the world to God," its successors elected to absorb "the world in God," which is "the work of German idealism." Idealism then surrendered to materialism, which adopted "an agnostic attitude before the problem of God." And with the onset of positivistic science God becomes unknowable (Zubiri 1981, 306).

This cursory glance at the history of systems of speculative reason exhibits both the introduction as well as the loss of God from the thought worlds of adherents to these systems. What is interesting to note, and which hopefully will be demonstrated during the course of this argument, is that there were metaphysical reasons for doing so.

For our current purposes, suffice it to say that there are authors who have uncovered a similarity between the tradition-boundedness of scholarship in the humanities and professional work in the sciences. This is a topic which will be more fully developed when discussing Thomas Kuhn; but for now we give our attention to the manner in which divergent or competing paradigms are encountered and organized in the discipline of theology. Wentzel van Huyssteen comments on how this process can be described in Kuhnian terms:

> What Kuhn called the incompatibility, and even the incommensurability, of paradigms is ... particularly noticeable in systematic theology. Theologizing within different paradigms or conceptual models, ... theologians ... might identify different problems and initiate different solutions while apparently working from the same basic commitment or premise. (van Huyssteen 1989, 65)

When faced with an option between competing alternatives of theory or model, the decision reached "is never ... entirely on rational or irrational grounds" (van Huyssteen 1989, 66). Rather, the selection describes "a conversion, a sudden gestalt switch that cannot be forced rationally or otherwise." As Kuhn judges this to obtain in the sciences, so van Huyssteen affirms that it is also true of rational activity in a broader sense, "which includes theology" (van Huyssteen 1989, 67).

Van Huyssteen next discusses the actual methodological similarities of the two disciplines and quotes Sallie McFague approvingly to this end:

> Scientific positivists have their colleagues in theology, for the assumption that it is possible to go directly from observation to theory without the critical use of models has its counterpart in those who assure it is possible to move from the story of Jesus to doctrine without the critical aid of metaphors and models. (van Huyssteen 1989, 3)

In opposition to this view, some respondents criticize that unlike the sciences "theological statements cannot be tested directly against their object" (van Huyssteen 1989, 94). Yet the same may also be claimed for certain scientific theories wherein the data in support of them do not derive from direct observation, such as scenarios for the early universe. Van Huyssteen, though, defends the theological enterprise by arguing that

> statements about God ... can nevertheless be evaluated indirectly in terms of our experiences of ourselves and the world. Not only does this indirect testing become the key to understanding the scientific rationality of theology, ... but it is also clear that faith and religious experience provide [us] with an ... empirical basis for theological science. (1989, 94)

It is not our purpose here to argue for or against theology as a strict science. What concerns us is the operational commonality of religion and science concerning how data, theories, models and traditions are dealt with. We may provisionally conclude that there seems to be a significant functional overlap in this respect. We will return to this theme once we have established its counterpart.

The Nature of Science in Religious Terms

> [S]cientific inductions can no more be proved by the Bible, than the doctrines of Christianity can be established by scientific investigation.[2]

The aim in this section is to outline the developments which bequeathed to science its current form. It will be noted that the contours of science were given shape by decisions themselves more religious-philosophical than scientific. To this end we will draw from the work of Thomas F. Torrance (b. 1913).

Torrance locates the watershed of the modern mindset in the first modern thinker, namely Rene Descartes (1596-1650), who argued that reality is dualistic, thereby dividing the "realm of material extension, essentially geometrical and mechanical in character, [from the] realm of unextended thinking substance" (Torrance 1984, 11). This rule for thinking effectively divorced the mind from the body; but what is more, Descartes' "application of mathematics to space and matter" ushered in "the mechanistic view of the universe," since for him the body is a machine. Sir Isaac Newton (1642-1727) was to follow Descartes' mechanistic lead but could not adequately reconcile "the role of God in regulating the chain of mechanical causes and coping with the emergencies resulting from irregularities within the universe" (Torrance 1984, 28). This difficulty gave rise to the notion that the God-world relation is conspicuous only in the gaps of scientific knowledge. With this understanding the greater the degree to which these lacunae are bridged, the more God becomes ousted from the system. If God occupies only the gaps, there will be ever-diminishing room for God to operate if the universe continues to yield "to mechanical explanation in accordance with Newton's laws" (Torrance 1984, 29). In the extreme, Laplace at the time of Napoleon "developed the Newtonian system of the world to its most complete form, in apparently demonstrating the internal stability of the solar system according to completely self-regulating and eternal law," driving him to maintain that he had no need of the God hypothesis.

Contrary to the Newtonian scheme, John Locke (1632-1704) spearheaded a departure from Cartesian rationalism toward an Aristotelian epistemology, already having been adopted by the theologian Thomas Aquinas, where "there is nothing in the mind which was not first in the senses" (Torrance 1984, 32) (even though Descartes also endorsed this notion). The empirical approach emphasized not the "objective orientation of appearances" as did Newton but "their subjective orientation: the way in which objects appear to the observer." By introducing a subjective element "of human knowing into science ... [Locke] laid the foundation for the inductive observationalism and empiricism of later times" (Torrance 1984, 34).

Such a mode of thinking, however, is also not without its difficulties. For instance, it was impossible to account "for mathematics on the basis of appearances alone." If extended to its logical conclusion, employing this method would lead to a Berkeleyan (1685-1753) form of idealism in which the existence of material substances could not be assumed on "observational or empirical grounds." One would then merely report on the impressions that are made in the minds of the beholders. Regrettably, "on Locke's theory of ideas, as Berkeley showed, the notion of material substances independent of our perception is meaningless" (Torrance 1984, 34).

David Hume (1711-76) was to add to the discussion by demonstrating that "not only do we have no empirical evidence for other minds, but we do not have [it] for the independent reality of our own" either (Torrance 1984, 35). An empiricist epistemology is problematic not only for science but for theology as well, since

> the negative conclusions of Hume's empirical positivism, the failure of the natural reason to establish an intelligible connection between the universe and God, also revealed the failure of this whole approach to account for the inherent intelligibility of the universe which made it accessible to rational scientific inquiry. (Torrance 1984, 35)

Immanuel Kant (1724-1804) was to prove monumental in providing a solution to the above difficulty. As Torrance explains,

> The importance of Kant in the development of European thought is located in his attempt to construct an epistemology in which the empirical contribution of Locke, Berkeley and Hume was combined with the more rationalist approach to knowledge of Descartes and Leibniz, in such a way as to show that purely theoretical *a priori* elements and empirical *a posteriori* elements operate together in all our knowing, in everyday life as well as scientific constructions. The result of this was his famous *synthetic a priori*. (Torrance 1984, 36)

A distinct turn thus took place in Enlightenment thinking, for autonomous reason shifted the center "of intelligibility to the human pole of the knowing relation" (Torrance 1984, 38). Unfortunately, this led to a paradox wherein the "natural order of things accessible to rational inquiry and formulation" and the grounds "upon which laws of nature depend is lodged in the mind and not in things themselves."

Hence Kant effectively undermined "the 'objectivity' of the phenomenal object" leaving observation as "not 'objectively given' but 'subjectively wrought'" (Torrance 1984, 39). The profoundly important implication, as we mentioned previously in relation to historical accounts, "was that scientific activity does not so much explore as shape nature, and does not so much discover as create reality" (Torrance 1984, 40). Kant's legacy is the attitude that "we can understand only what we can construct for ourselves," and the inheritance we have received from him is the view that "an object *is* what it becomes in our cognition of it." All post-Kantian science must therefore contend with the realization that "in all our knowing there is a real interplay between what we know and our knowing of it" (Torrance 1984, 42).

Enter the Philosophy of Science

The word theory *has the same Greek root as* theatre. *Both are concerned with putting on a show. A theory in science is no more than what seems to its author a plausible way of dressing up the facts and presenting them to the audience.*[3]

Having sketched the history of trends in modern philosophy from Descartes to Kant, we now embark on what may be referred to broadly as the philosophy of science. Regardless of whether it has been their intent, it can be found in the foreground of the successors to Newton that to subscribe to an empiricist framework means the expulsion of metaphysics, since for them "fundamental ideas of a philosophical or epistemological kind can have no place in natural science" (Torrance 1984, xi). The trouble with a viewpoint of this sort is that it cloaks a thinly disguised dogmatic rationalism which presupposes that the facts of observation are brute ones "without any intelligible component, and that ideas are abstracted from them by way of logical processes alone." Since the time of Auguste Comte (1798-1857) this method of relating to the world came to be known as *positivism*. Albert Einstein (1879-1955), though, was one scientist who insisted "that there is no logical bridge between ideas and empirical realities." Instead, "empirical and theoretical factors inhere in one another," and allowances should be made for this in our epistemology.

Rationalistic empiricism thus gave rise to a "lapse in the concept of the inherent intelligibility of nature" and in turn fed the instrumentalist science of the modern era (Torrance 1984, 6), where scientific models are assumed not to describe reality but are useful only as tools for understanding the world. The mentality which these views spawned can be expressed in this fashion: "it is we who, by our rational and scientific operations, clothe the universe around us with form and structure and thereby give it meaning for ourselves." But the problems for positivism multiply, since detachment becomes an unattainable goal, for "positivism is actually attached to definite presuppositions which it will not allow to be questioned" (Torrance 1984, 200). This is precisely the position that has been so radically undercut in contemporary philosophy of science. All knowing and hence "all science operates with basic beliefs" that are not subject to positivistic logic. All epistemological activity, it could be argued, functions *worldviewishly*, and the major figures who brought this to light in the twentieth century will now be presented.

The Recent Situation

> *[Whereas natural laws describe what happens,] [a]ll that they prescribe are our expectations.* (MacKay 1974, 31)

To begin with, Sir Karl R. Popper (1902-94) asserted that the idea "of absolutely certain, demonstrable knowledge—has proved to be an idol," and that all scientific utterances have a provisional character (Flew 1984, 44). For Popper the hallmark of science is not its possession of truth but its quest for it, intimating that not verifiability but falsifiability bears the actual mark of true science. This replacement which Popper made seems to be a more modest one about the powers and limitations of the scientific enterprise. He admits the interpretive nature of information derived from observation. Brute facts thereby become preinterpreted "in the light of either clearly formulated or subconsciously entertained theories" (van Huyssteen 1989, 27). Theory-formation involves the agenda of the theory-former, namely the observer's perspective or "angle of observation" together with "theoretic interests." What lies between observation and theory and what determines a scientist's point of view are the entire constellation of investigations s/he has selected as well as the host of "conjectures and anticipations" which engage him or her. This "total spectrum of expectations" opposes the idea of pure objectivity, and instead of "naked observation" is substituted its "theory-ladenness."

Next Thomas Kuhn (1922-96) made significant strides in the landscape of philosophy of science. He suggested that "something like a paradigm is prerequisite to perception itself" (Kuhn 1970, 113). Hence observation depends on philosophical factors such as what a scientist's "previous visual-conceptual experience has taught him [or her] to see." Even choices surrounding which equipment to utilize "carries an assumption that only certain sorts of circumstances will arise" (Kuhn 1970, 59) or be allowed to count. Kuhn believes that the history of science has been punctuated by periods of paradigm-adoption, followed by what its practitioners might regard as "normal science." As anomalies build up, that refuse to conform to the existing scheme, a critical mass is reached and the discontinuity Kuhn refers to as a scientific revolution is kindled. At this point there is an upheaval, and new paradigms compete to explain the new trends in an attempt to become the newly accepted paradigm which will once again usher in a time of normal science. Kuhn also introduces the concept of the instability of data during revolutionary periods. The same data can no longer be approached in the same way, which then calls for a new perspectival structure. For example, in the contemporary scene a "pendulum is not a falling stone," as it was to the Greeks, "nor is oxygen dephlogisticated air" (Kuhn 1970, 121). New paradigms must appear which not only contain the new categories but afford the language enabling us to speak of the data in renewed terms. In the face of this eventuality, Kuhn cautions that, "it is hard to make nature fit a paradigm."

Kuhn takes additional issue with Popper's idea of falsifiability. For something to be capable of falsification, Kuhn contends that hypotheses must be able to surrender to empirical testing and then be found wanting. Kuhn's assertion, according to Philip E. Johnson, is that "[t]he problem with this criterion is that it is impossible to test every important scientific proposition in isolation" (Johnson 1991, 120). "Background assumptions" in particular need to be made plain in order for testing of scientific statements to proceed, and these are embedded in the reigning paradigm. Paradigms color the way we view the world and dictate how we see. And anomalies by themselves can never contravene a paradigm, as Johnson reads Kuhn, since "its defenders can resort to *ad hoc* hypotheses to accommodate any potentially disconfirming evidence." Not only is it difficult to mold nature into the shape of a specific paradigm, it is also a formidable task to dislodge a paradigm once it has become firmly entrenched and captured the imagination of its adherents.

With the establishment of a new paradigm and the new perspective it generates there is a reorientation so as to assimilate the anomalous. With any paradigm "people tend to see what they have been trained to see," implying that these persons are "prone to paradigm-influenced misperception" (Johnson 1991, 120-1). All paradigms operate as corrective lenses on the world which actually distorts one's vision. For instance, in the crumbling of the Ptolemaic cosmology, certain "celestial phenomena ... were not 'seen' until the new astronomical paradigm of Copernicus legitimated their existence." Hence paradigms also have a social role to play in that the communal agreement to adhere to a paradigm functions as a legitimating structure for the activities occurring within that group. "[P]erception is an interpretive process" (Kuhn 1970, 195), and to this end we have never viewed the universe in the same way after the work of Copernicus received general acceptance. Membership in what has been adopted by convention validates one's work.

Scientific revolutions for Kuhn follow crises which give way to paradigm shifts that in turn herald a new period of normal science (van Huyssteen 1989, 50). The newly accepted framework "does not merely propose different answers to the questions scientists have been asking, or explain the facts differently; it suggests entirely different questions and different factual possibilities" (Johnson 1991, 121). This prompts Kuhn to evaluate opposing paradigms as incommensurable, intending by this to convey the thought that the difficulties groups of scientists experience in communicating intelligibly with each other are due to their employing disparate paradigms. The terms of one paradigm are not easily translatable into those of another, leading Kuhn to submit that for such communication to be successful "one group or the other must experience the conversion that we have been calling a paradigm shift" (Kuhn 1970, 150). For Kuhn, this type of transition "cannot be made a step at a time, forced by logic and neutral experience. Like a gestalt switch, it must occur all at once (though not necessarily in an instant) or not at all."

The selection procedure of a candidate for a new paradigm frequently takes place on the basis of personal and aesthetic considerations (Kuhn 1970, 155-6), and a decision in favor of that choice on the part of individual scientists is often based on belief (Kuhn 1970, 158). As paradigms correspond to traditions, the views of Kuhn are found to boast "wide applicability."[4] His insights fit not only scientific perspectives but also "a broad philosophical outlook" (Johnson 1991, 121). Kuhn's strategy reveals that, contrary to positivism, the sciences share much in common with other disciplines: "[i]n fact, what we have here is a definite relativization of [positivistic] rigidity" (van Huyssteen 1989, 62). Science and theology may each be judged as a "sociohistorically determined conceptual model" which is "dominated by the impact of paradigms."

The terms "determined" and "dominated," however, might be too strong. Kuhn's work is a needed corrective to the logical positivism of the 1920s and 1930s, but his ideas reside at the other extreme and thus describe a pendulum swing in the opposite direction. Perhaps this tactic was warranted at a time when his point needed to be made, yet in the intervening time the blow has been softened somewhat.

The final thinker in the area of philosophy of science to whom we direct our attention in this section is Michael Polanyi (1891-1976), who discusses the role of the scientist in the endeavors of science. He asserts that "far from being neutral at heart, [the scientist is] passionately interested in the outcome of the [scientific] procedure" (Polanyi 1946, 38-9). What is more, the scientist also reaches decisions as to "what weight to attach to any given set of observations in support or refutation of a theory on which they seem to be bearing" in a fashion that is "[u]ltimately ... a matter for his personal judgment" (Polanyi 1946, 93-4). And the personal judgments do not stop there but rest with every observer, since humans have a "personal stake in all [they know]."[5] Polanyi continues in this same vein:

> The participation of the knower in shaping his [or her] knowledge, which had hitherto been tolerated only as a flaw—a shortcoming to be eliminated from perfect knowledge—is now recognized as the true guide and master of our cognitive powers.... We must learn to accept as our ideal a knowledge that is manifestly personal. (Polanyi 1959, 26-7)

Polanyi further argues that in terms of human knowledge "we can know more than we can tell" (Polanyi 1967, 4). To this end he offers the illustration of our recognition of someone's facial features. We are able to acknowledge the face of someone we know, but "most of this knowledge cannot be put into words." Polanyi refers to this type of knowing as the "'tacit dimension' of consciousness, an underlying knowledge so taken for granted, so obvious as to become invisible to us" (Owens 1983, 116). It remains part of the human condition to "operate with a 'subsidiary awareness' and an implicit knowledge on which we rely in all our explicit operations" (Torrance 1984, 112). This dimension functions "through an unaccountable intuitive apprehension of a structure" in a person's world of experience. What must be emphasized here is that this knowing is operative not only within the arts but also the sciences. Indeed the contributions of Popper, Kuhn and Polanyi can be applied to the humanities as well as to science.

A Diagnosis of the Enduring Problem

> [T]he faith which the positivists displayed in natural science was not itself arrived at scientifically.
>
> Perception is not insulated from theory. Theories cart along their own confirmations.
>
> Warrant will have to be relative to a body of beliefs. It cannot be relative to a body of certitudes.[6]

Despite the attempts by proponents of science in its positivistic mode to delete human factors from its framework, it is precisely the human elements, for philosophers such as Kuhn, which enable science as ordinarily understood to proceed. Human factors are neither eradicable from scientific activity, nor is this even a desirable outcome. Reinforcing this claim, one author, himself a scientist, issues the following decree:

> The rational part of research would, in fact, be useless if it were not complemented by the intuition that gives scientists new insights and makes them creative. These insights tend to come suddenly and, characteristically, not when sitting at a desk working out the equations, but when relaxing, in the bath, during a walk in the woods, on the beach, etc. During these periods of relaxation after concentrated intellectual activity, the intuitive mind seems to take over and can produce the sudden clarifying insights which give so much joy and delight to scientific research. (Capra 1983, 39)

As has already been established, the move from observation to theory-formation includes inductive steps, namely interpretation of data and the systematization of information. These "are products of human invention and insight" (Ratzsch 1986, 23) and do not automatically spring from the objects investigated. Theory-formation is not a rigorous undertaking, nor is there a "logic of discovery" (Ratzsch 1986, 23-4). Instead, the leap from observation to theory involves imagination—a distinctly human characteristic. "[T]heories may be suggested by the data, [but] they are not logical consequences of the data. They are rather the results of creative insight on the part of humans" (Ratzsch 1986, 24-5).

Kuhn believes he has demonstrated that "human subjectivity (in the form of values of the community of scientists)" occupies a central role in science, making it "a decidedly human pursuit" (Ratzsch 1986, 55). Science is able neither to legitimate its own methodology nor its presuppositional foundations, allowing us to conclude that science is not omnicompetent (Ratzsch 1986, 98). Van Huyssteen cites Wolfhart Pannenberg who claims that the pronouncements of science are "ultimately founded on general world views of a profoundly philosophical and/or religious nature" (van Huyssteen 1989, 80). These ultimate beliefs usually operate on the unconscious level and escape the attention of scientific practitioners (Torrance 1984, 199). Hence in so far as both science and religion participate in assuming fundamentals, "integrating and giving meaning to available data," testing hypotheses and "providing solutions to puzzles" (Van Huyssteen 1989, 80-1), they are methodologically and epistemologically related (Torrance 1984, xii).

Theories continue to shape the way the world is viewed (Gerhart & Russell 1984, 10), yet science still gives priority to "theoretical understandings ... over the naive understanding of our immediate experience" (Gerhart & Russell 1984, 38), and persists in appraising disciplines "as more or less 'mature' in terms of the extent to which they make use of analysis" (Gerhart & Russell 1984, 41). According to Torrance, though, rigid scientific methodology has been deemphasized recently in favor of integration, which has "opened up the way for profounder conceptions of systems and organization, so that justice may be done to the multivariable, many-layered character of the universe" (Torrance 1984, 87). Indeed the human capacity to understand the world reflects nature's understandability. Torrance makes the following affirmation:

> This fact, that our concepts and their proper interconnections are intuitively related to and controlled by nature or the real world, implies an astonishing harmony or correlation between its inherent comprehensibility and the structure of our human comprehending; the fact that the world of our sense experience is comprehensible to us, said Einstein, is a miracle. (Torrance 1984, 79)

For Torrance there is a place for wonder in scientific inquiry, especially when we meet up with the unusual, for then we would be prepared to respond to it not with the fixed canons of stereotypical categories but with an openness to new possibilities (Torrance 1984, 77-8).

If we are to believe these authors and their sentiments, science possesses a religious dimension since it bears a human stamp (Gerhart & Russell 1984, 77). In addition to the wonder Einstein and Torrance spoke about, theories have also been lauded because of their elegance, which constitutes another extra-scientific category. Beauty too can "give rise to a conviction that can amount to faith.... The beauty of a theory is persuasive of its truth. But the question of beauty is a limit-question, a value question, and therefore evidence of a religious dimension in science."

If there is such a similarity between the enterprises of science and religion, then how has so much friction arisen between them and how can this be rectified? A tentative explanation surrounds the idea of a clash of fundamentals, namely the worldview of materialism versus one with a transcendental component. This proposal will be worked out below, but for now let us observe how the two interact. C. Stephen Evans contends that there need not be an atmosphere of tension between religion and science, largely because the two normally do not "give competing explanations on the same level for ordinary events" (Evans 1982, 125). The trouble manifests itself when science regards its assertions as "ultimate and final," for this appears not "to be a testable scientific claim." Questions of ultimacy "cannot be answered by science itself." Religion, on the other hand, undertakes to comment precisely on "why nature exists at all, and why nature has the ordinary characteristics which science investigates." For Evans, religion and science are not "essential rivals" because each commands its own sphere of competence. Evans understands the two as offering explanations which "are not of the same type or on the same level." Science deals with "what" and "how" questions; religion with "why."

Offsetting Evans' either-or stance, Vincent Brummer submits a both-and characterization by stating that despite functioning at different levels, religion and science are mutually dependent. He is persuaded that "religion is dependent on science in order to know the changing factual environment of which it has to make sense," and science "is dependent on the conceptual models of a religion or view of life in order to determine its own sense or meaning" (Brummer 1991, 12). In this outlook, the interaction of science and religion may be congenial, and conflict need not become inevitable. Nevertheless, the problems remain. Advancements in the field may be made with the contribution of Langdon Gilkey (b. 1919), to whose thought we now turn.

The World According to Gilkey

> *A complete, consistent, unified theory [of physics] is only the first step: our goal is a complete understanding of the events around us, and of our own existence.* (Hawking 1988, 169)

> *[I]f we do discover a complete theory, it should in time be understandable in broad principle by everyone, not just a few scientists. Then we shall all, philosophers, scientists, and just ordinary people, be able to take part in the discussion of the question of why it is that we and the universe exist. If we find the answer to that, it would be the ultimate triumph of human reason—for then we would know the mind of God.* (Hawking 1988, 175)

While we must put aside for the moment the issue of God's mind, the presence of limit-questions which science poses reveals a basic human need. Humans do not appear to be satisfied with scientific descriptions of "what" and "how" but press on toward speculations as to "why." Does this driving force indicate the area where religion and science meet and overlap, or where science passes the baton to religion/philosophy, or is there a third and more suitable description of their potential interaction? Langdon Gilkey makes some headway in this respect.

Already in his earliest book Gilkey offers the following account as to the relationship of truth to science and religion:

> In so far, ... as religious concepts are taken to represent literal truths, they are scientifically and philosophically untrue, and religiously irrelevant. Paradoxically, a myth can only be true as a religious affirmation, if it is untrue as a literal description of fact. As literal truths, myths are "prescientific," and must be discarded—but it is precisely at this point that they have no relevance for religion. (Gilkey 1965, 348)

In Gilkey's estimation, science functions as a "cultural vision ... in shaping, preserving, and guiding the community" (Gilkey 1979, 30). But the story does not end there, for science exhibits religious contours by entertaining elements of "ultimacy and sacrality, in claiming to embody an ultimately valid view of reality." Whereas it would be overstepping the bounds of science and religion if each were to address the concerns of the other, to the degree, for instance, that the doctrine of evolution answers "questions of ultimate origins and destiny," it then becomes a myth, as Stephen Toulmin insists (Gilkey 1981, 38-9). If this is a task that evolutionists in fact undertake, then it could be argued that the line has been crossed, since information about ultimate origins and final destiny is not something that evolutionary theory can yield, especially if natural selection truly operates randomly. If the ultimate fate of organisms can be mapped out, barring of course a nuclear holocaust or collisions with extraterrestrial objects, then natural selection is a determined and not a random proposition. There will be more to say about this in the chapters ahead.

If myths can be located in science, however, they are neither scientific nor precise (Gilkey 1981, 39). They do not arise from scientific procedure and they are not quantifiable or reducible to mathematical formalisms. Instead myths develop from the presuppositions of a system of thought. Despite the rationality of the discipline, science is unable to assess its own presuppositions rationally (Gilkey 1981, 42), since "[k]nowing is ... through and through a human act" (Gilkey 1981, 47), wherein convictions are involved. Gilkey cites Polanyi approvingly to this end: "Every act of factual knowing has the structure of a commitment" (Gilkey 1981, 155).

In Gilkey's appraisal, the proper role of theology is "exploring the meaning and the validity of mythical language, or the symbol system of a given religious tradition" (Gilkey 1981, 107). Yet theology is not the only discipline to contain mythical language. Science too displays "a dimension of ultimacy" comprehensible "through categories similar to those we have called 'mythical'" (Gilkey 1981, 124). Gilkey does not judge this as a shortcoming; rather, the drawback rests with science's attempt to unravel the mysteries of the universe without reference to mystery. Science does not lack its metaphysical moments—only the awareness or admission of their occurrence. For those occasions Gilkey warns that "[i]f a sentence bred in science's neatly fenced pasture wanders off to frolic in metaphysic's Elysian fields, it had better expect to receive new brand markings" (Gilkey 1981, 152).

Regrettably, this does not reflect the attitude of contemporary science, for there is a dogmatic edge to the claim that "scientific knowledge represents the *sole* cognitive entrance into what is real, ... Nature, therefore, is as science defines it, and reality is equivalent to nature as determined by science" (Gilkey 1989, 285). Contrary to this mindset on the part of certain scientific practitioners, Gilkey contends that observation as "observer-dependent" points to both the inadequacy and relativity of our knowledge, indicative of a limit that is "essential and not ... accidental" (Gilkey 1989, 287).

Inquiry for Gilkey involves as much "intuition, imagination, and ... insight, as it does logical brilliance," and admits as much art as it does science (Gilkey 1989, 289). Hence Gilkey evaluates science as another arena for the sacred, by which he intends "the unity of reality, meaning and value" (Gilkey 1989, 297). To Gilkey's thinking, intuitions "represent aspects or glimpses of the sacredness of reality" as well as "traces of the divine." Stephen Toulmin coined the term "limit questions" which "appear at the boundary of a discipline," sometimes marking its basic assumptions, and are thus incapable of being addressed from within it (Gilkey 1985, 257-8). Gilkey calculates that these questions "raised by science, but ... beyond [its] reach," afford "hints of meaning and of value" that "call for philosophical and religious reflection" (Gilkey 1989, 297). And together they constitute signs or "disclosures of the sacred," the presence of which is "characteristic of reality" (Gilkey 1989, 298).

The Nature of Nature

> *[As physicist Max Planck declared] Science cannot solve the ultimate mystery of Nature. And it is because in the last analysis we ourselves are part of the mystery we are trying to solve.* (Barrow & Tipler 1988, 123)

Having discussed the nature of and the interplay between religion and science, we now turn to an examination of these presuppositions of science using geology as a specific example. Science prides itself on its methodological standards of testability, repeatability, quantifiability, predictability and control, thereby promoting the division between the public sphere of science and the private realm of ethics and religion. In so doing, science believes it occupies the sphere of the objective whereas the realm of everything else is subjective in character. Cracks in the scientific edifice, though, periodically surface and when they do, science's extrascientific features become detectable. Scientific objectivity as more alleged than real is particularly evident in the assumptions surrounding the theory of evolution. We begin, however, with a notable illustration drawn from the field of geology.

A difficulty arises in the attempt to describe the earth's history prior to the time of observers who could report on it. Reijer Hooykaas specifies four approaches to formulating such accounts, the two extreme positions being the most important for our purposes. The first, at one end, is termed *catastrophism*, where the past geological changes of the earth are believed to be both qualitatively and quantitatively different from current ones (Hooykaas 1969, 3). That is, these past upheavals not only occurred at a different rate but were also more vigorous. The two mediating stances emphasize either the qualitative or quantitative differences alone and often go under the name of *actualistic gradualism* or a variant thereof (Hooykaas 1969, 4). The fourth, at the opposite end, is referred to as *uniformitarianism*, where past forces differ neither in rate nor intensity from present ones (Hooykaas 1969, 14). This last type is the one adopted by most in the contemporary scientific community, since in their view "the present is the key to the past" (Hooykaas 1969, 17).

For uniformitarianism, geological forces undergo no modification over time, and conveniently so, since otherwise our study of the geological past would be hampered by cognitively inaccessible rates and intensities. Yet this is a far-reaching claim. One implication of the uniformitarian doctrine is that notions of progression are excluded from its purview (Hooykaas 1969, 101). In this sense uniformitarianism is "a-historical" for when "it admits a development in a certain direction, it gives up its fundamental assumption."

Catastrophism is often rejected because of its religious overtones, mainly stemming from an acceptance of the Genesis Deluge and its supposed explanatory power, but it is not the only position to convey religious sentiments. Similarly, problems occur whenever scientific principles are accorded the status of laws (Hooykaas 1969, 164). In this regard, "[t]he *principle* of causality has been mistaken for a *law* of causality," and in like manner "the principle of uniformity sometimes has become a *law* of uniformity." The situation which prevails seems to be that once a methodology "has been made the touchstone for scientific reasoning," it then takes up residence at the metaphysical level. With this eventuality, "[t]here is great danger that a methodological attitude may degenerate into narrow dogmatism" (Hooykaas 1969, 40). If so, it becomes an impediment in the advancement of science, which instead may ossify.

Descriptions and expectations of nature are grids or lenses that are placed between observing subjects and observed objects. The trouble is, nature might not always conform to the framework we have placed upon it. Our map which we thought would supply us with familiar directions has led us to the edge of the page. Does this mark a point where an as yet largely unopened map is warranted?

Before we venture out onto uncharted territory, it would be best to acquaint ourselves with the borders of our current region. This calls for an excursion into the sciences in order that we might be conversant about what lies at the boundary between established science and speculative philosophy.

CHAPTER TWO

The New Physics Applied to the Old Nature

Evolution does not necessarily select for truth of conceptualizations. (Ratzsch 1986, 125)

Matter itself is far from understood; far less, then, is it capable of being the paradigm of explanation by which everything else can be understood. (Ferre 1993, 164)

RECENT APPROACHES from within conventional science have prompted some scientists to broach philosophical topics. As we focus our attention on the twentieth century revolution in physics we will determine potential epistemological and metaphysical implications that may be drawn from it. What undoubtedly will become clear is that contemporary physics is as much speculation as it is observation.

The Quantum World

The initial stop on our tour is quantum mechanical theory and how it fosters a new look at material reality. A quantum is defined as the energy released when subatomic particles move from one energy level to another. This energy comes in discrete packets known as *quanta*. What is interesting to note about the movements between energy levels generating these quanta is that there are no intermediate steps involved. A particle, say an electron, is at one orbital or shell at one point and then at another with seemingly no interval of space or time. And even the notion of "particle" is a specious one since it exhibits just as many wave-like characteristics. Lacking both the capacity to observe a subatomic entity as well as the justification for presuming its particulate nature, scientists are forced to model the item in mathematical terms for only through this avenue can one speak about the nuclear world as actually "made up of" something (Davies 1983, 162).

Some subnuclear entities apparently display the ability even to move in and out of existence, which yields a curious type of "quantum logic." As Paul Davies explains in reference to the law of identity in philosophy, "the rule that something cannot both be and not be such-and-such is dropped. The motivation for this is that in quantum physics the notion of 'to be' is more subtle than in everyday experience" (Davies 1992, 26). In his view it is perhaps more felicitous to think in terms of there being "no elementary particles at all," because it would seem that every subnuclear particle "is made up of every other. No particle is elementary or primitive, but each contains something of the identity of all the others" (Davies 1983, 48).

Quantum mechanics, while admittedly mechanical to a degree, applies mostly to the subatomic level; whereas at the macroscopic level "[q]uantum effects are usually negligible" (Davies 1992, 61). This has important ramifications on the cosmological scale. The currently accepted model for the beginning of the universe, though historically not without its competitors, remains the Big Bang, where in approaching the time zero "the universe was compressed to minute dimensions" and here only quantum or what is known as string theory would apply (Davies 1992, 62). Science sometimes lacks the clarity as to how to proceed when interpreting quantum effects, and their significance is still being worked out; but what does come across as patently clear is that "the reality exposed by a quantum measurement is decided in part by the questions the experimenter puts to nature" (Davies 1992, 185). Davies concludes that "[t]he world is ... neither wholly determined nor arbitrary but ... an intimate amalgam of chance and choice." What compounds the enigma is that "the quantum wave is not like any other sort of wave anybody has ever encountered. It is not a wave of any substance or physical stuff, but a wave of knowledge and information. It is a wave that tells us what can be known about the atom, not a wave of the atom itself" (Davies 1983, 107).

The Relativistic World

Owing to the work of Einstein, we in the modern Western world can no longer look upon the universe in the same way as before his ground-breaking insights. Einstein's special theory of relativity demonstrated that there is no absolute space or time. Both become relative to one's frame of reference and together form a spacetime continuum. An outgrowth of this discovery is that matter is now understood as a form of energy and vice versa: "Matter is compressed and relatively stable energy, and energy is a dispersed, excited state of matter" (Owens 1983, 83). And regarding Einstein's general theory, which deals with gravity and accelerated frames of reference, John Wheeler comments on the nature of the relation between space and matter in these terms: "Space tells matter how to move, and matter tells space how to curve" (Barbour 1990, 110).

Einstein's contributions represent the culmination of the Newtonian mechanistic world-picture. In Einstein's perception there remains a one-to-one correspondence of cause and effect in the material world. According to Jeremy Bernstein, "Einstein's space-time consists of points whose positions and times are to be determined by classical procedures using rulers and clocks of a kind that any nineteenth century physicist would be comfortable with" (Bernstein 1973, 154). Einstein was driven by the assumption that God does not play dice with the world, meaning all events could be determined, provided that the information from all preceding causal factors could be secured. This is precisely what the Newtonian worldview holds: relativity does not imply relativism. Einstein believed in a dependable universe that bears a regular structure and in which there are no surprises or discontinuities.

Such a conception of the universe differs markedly from the quantum outlook. The quantum approach brought with it aspects of nature that were no longer to be understood in fixed or absolute terms. A photon is the quantum of electromagnetic radiation and an electron is the quantum of electric charge (Owens 1983, 83). These and other subatomic entities behave like waves as much as particles. Whereas relativity emphasizes continuity, determinism and predictability, quantum mechanics offers instead discontinuity, indeterminism and unpredictability. More on this presently.

Einstein is credited with originating the general theory of relativity, "[o]ne of the main features of [which] is that the affairs of matter cannot be separated from the affairs of space and time" (Davies 1992, 48). The big bang model of the cosmos is itself an outgrowth of this theory. If space and time form a continuum, as his theory requires, then time was generated at the same instant as was space. Both space and time had the same beginning. This is why it does not make sense to ask what occurred prior to the big bang, since time is necessary for an event to occur. With no time there are no events, at least as we know them. In line with this reasoning, "cause and effect are temporal concepts" and are thus inapplicable to a "pre"-big bang state (Davies 1983, 39).

The beginning of the universe on this view is a point of infinite material density and gravitational force and is termed a "singularity" (Davies 1992, 49). As time progressed from this point it revealed a discrepancy between its natures on the micro versus macro scales. On the macroscopic level we perceive an "arrow of time"—a definite direction to the course of events. Our experience is usually one of time asymmetry where effects can never be reconverted into the state they assumed immediately prior to the causes acting upon them, no matter how much energy is injected into the system. The movement of events is unidirectional and irreversible; anyone having broken a glass object can attest to this. Yet at the microscopic level "[t]he collision between any two molecules is completely reversible and displays no preferred past-

36   *God and the New Metaphysics*

future orientation" (Davies 1983, 126). That this elasticity should have developed the way it did is another mystery.

The initial point singularity gave birth not only to time and space but to natural laws as well. As was mentioned previously, the laws of nature are descriptions of patterns of material behavior and as such can be formulated only in the context of space and time which contain the material so described. If this is the case, then "the laws of physics must break down at the singularity" (Davies 1992, 50). This drives one to pose the question as to the formation of these behavior patterns. Did the laws describing them develop over time as indeed time and space did? And if they evolved to this point could they evolve further still?

In more general terms, does the universe form a closed system "containing the reason for its existence entirely within itself?" (Davies 1992, 68) With the laws already in place the universe can cope on its own and unfold quite nicely. But where and how do they originate? One solution is to opt for the transcendence of natural laws, where they once occupied a realm different from our own and were deposited here at some point. If one decides, however, that "the laws are not transcendent, one is obliged to accept as a brute fact that the universe is simply *there*, as a package, ... [with] the laws built in" (Davies 1992, 92).

It might be hasty to suggest that these are the only alternatives. Another approach is to treat space and time and the laws that describe them "as fields which themselves simply 'froze' out of a primeval soup of pregeometrical elements" (Davies 1983, 160). In this proposal the universe would have been formed in the manner of "complexity frozen out of simplicity." At the present moment, though, this notion is little more than conjecture.

Einstein Versus Bohr

Our next stopping point is a consideration of the epistemological issues raised by the Einstein—(Niels) Bohr (1885-1962) debate. We begin with recalling the theme of waves alluded to earlier. For our present purposes we cite an experiment involving the firing of an electron toward a photographic plate. Upon being fired it has a probability of contacting the plate at a number of different spots, meaning the mathematical formalism known as a "wave function is spread out over a wide area" (Barrow & Tipler 1988, 459). As soon as the electron hits the plate, however, the wave function collapses and the probability of the electron hitting that spot becomes one and is rendered zero everywhere else. We mention this because a parallel exists between probabilities and knowledge. Once the electron makes contact with the plate there is no longer a need to speak of probability, for certainty regarding an electron's whereabouts has replaced it. So too our lack of precision in knowing the system becomes definite at this point.

Upon noticing this, Werner Heisenberg drafted his famous uncertainty principle. It rests on the ultimate dilemma that one cannot with accuracy measure and obtain information about both the position and momentum of an electron simultaneously. In order for readings to register on laboratory instruments and for scientists to read data from their equipment, a minimum of energy, say one photon, needs to be expended or invested in the system. The photon—a particle-wave of light—collides with an object, here an electron, bounces off and is caught in the instrumentation where it delivers its message, enabling a measurement/reading to be taken. Yet the photon in its trajectory carries with it sufficient energy to knock an electron off course. The reading that one obtains of an electron is then by virtue of bumping into it. By interfering with it in this way we do not observe its true velocity or direction. Heisenberg seized upon this idea and claimed that "nature is inherently unpredictable" (Davies 1983, 137). There is thus a fundamentally irreducible indeterminism at the world's foundation. As one undertakes to "reduce the level of uncertainty of" one or the other of position and momentum, this "serves to increase the uncertainty of the other" (Davies 1992, 30). Uncertainty is therefore "intrinsic to nature," an important implication of which is that the present does not determine the future exactly. This is evident in situations where "events occur that have no well-defined cause" (Davies 1983, 137), such as which nucleus in a radioactive substance will be the next to decay? The disintegration of a nucleus is a matter of probability where no one nucleus is a more likely candidate than another to undergo this random selection process. For quantum systems at least, then, "[t]he universe really is indeterministic at its most basic level" (Davies 1992, 31).

Niels Bohr of Copenhagen used this type of interpretation to suggest that the indeterminism in the physical universe was essential (Barrow & Tipler 1988, 460). "The observer and the world [are] so inextricably connected," he urged, that "many physical properties of atomic particles did not even exist before the act of observation [which] was necessary to bring these properties into existence." In light of this uncertainty, Bohr endorsed

> the empiricist principle that what cannot be measured, even in principle, cannot be said to exist. Bohr therefore denied reality to the notions of electron position and electron momentum prior to their measurement ... [they] would be determined by the particular experimental arrangement which the observer chose to interact with. (Barrow & Tipler 1988, 461)

Einstein took the opposing view. His stance was one of realism, and he along with two colleagues sought to counter the empiricist posture by using the Einstein-(Boris) Podolsky-(Nathan) Rosen or EPR thought-experiment. According to this idea, when two electrons are coupled on the same axis they reveal a total spin of zero—say one clockwise and the other counterclockwise. If the experimenter were to distance one of the electrons from the other while they remained on the same axis, say one light year apart, and then measure the one left behind, one would obtain a measurement of either clockwise or counterclockwise. The significance of this investigation is that while an experiment is performed on just one electron, information about the other is also obtained indirectly and one light year before it could usually have been secured. This drove Einstein to submit that the second electron has physical reality even prior to being measured. Such a finding would contradict Bohr's assessment since it is "impossible for the measurement on [the first electron] to bring into existence the properties of [the second]" (Barrow & Tipler

The New Physics Applied to the Old Nature 39

1988, 462). The first electron could not affect the second prior to one light year having elapsed.

The EPR experiment illuminates the counterintuitive nature of Bohr's understanding of quantum mechanics, namely "[t]he idea that the act of observation must have a non-negligible effect on the object being observed" (Barrow & Tipler 1988, 463). Einstein and colleagues appear to have provisionally overcome these difficulties, yet noteworthy at this juncture is the fact that information from the second electron is not observed but inferred. This injects into the deliberations an interesting question as to whether there can be a transaction of information at a rate faster than the speed of light, Einstein's lone absolute, or in other words, is there room for "instantaneous action at a distance"?

Not to be outdone, Bohr's response to this challenge was such that "[e]ven though there is no direct interaction between electrons number one and number two during the measurement, they are bounded together by the observer's decision to obtain information about electron number two by measuring a property of electron number one" (Barrow & Tipler 1988, 464). Since "the two-particle system is an indivisible whole, ... the system cannot be analyzed in terms of independent parts" (Capra 1983, 345). Bohr retained his view and the debate continued unabated with Einstein's efforts to refine his experimental design so as to circumvent Bohr's objections.

Erwin Schrodinger (1887-1961) next inserted into the proceedings his famous "Cat Paradox" in an attempt to offer a macro-scale illustration of a micro- or quantum world event. In his thought-experiment a

> cat is sealed in a steel chamber. If a radioactive atom decays within an hour, a hammer shatters a flask of hydrocyanic acid and the cat dies. If no atom decays, the flask is not shattered and the cat lives. After one hour the cat's wave function is a superposition of two states, ... the cat is both alive and dead.

For Bohr, the cat exists as a superposition of two states—both dead and alive—until such time as the system is interfered with.

In contrast, for Schrodinger the cat could be only one or the other but not both. Yet Bohr insists that opening the chamber to observe the contents has the effect of collapsing the wave function. (Probabilities are always reckoned as fractions, but once the above result is obtained, one possibility becomes one and the other zero.)

A similar experiment involves one ball and two boxes. One need only look in one box to determine the contents of the second and thus the location of the ball. "However, according to quantum mechanics the ball is, so to speak, half in one box and half in the other—just as Schrodinger's Cat is a mixture of dead and alive states before the chamber is opened—and suddenly 'materializes' in one or the other box at the instant of measurement" (Barrow & Tipler 1988, 466). The precise point where the wave function can be said to collapse, though, is problematic. For example, in the cat experiment "it is unclear who should be called the observer: is it the Geiger counter, the cat, or the human observer?" (Barrow & Tipler 1988, 467). Eugene Wigner's solution is that "the interaction of human consciousness with the physical system ... is responsible for the collapse." The issue remains unsettled as long as there are realists in the tradition of Einstein to engage in battle with the successors to Bohr.

At this point in the debate the indeterminism of the Copenhagen school seems to have won the day over Einsteinian determinism. On the micro-scale "[e]vents without causes, ghost images, reality triggered only by observation—all must apparently be accepted on the experimental evidence" (Davies 1983, 106-7). The new physics sides, at least tentatively, with the perspective that uncertainty is inherent in nature, leaving "the common sense view of the world, in terms of objects that really exist 'out there' independently of our observations," at a distinct disadvantage. John Wheeler even goes so far as to argue that since conscious observers participate in our perception of reality, "mind can be made responsible for the retroactive creation of reality" (Davies 1983, 111). Whether or not this is overstating the case, some researchers are convinced that "[b]y blurring the distinction between subject and object, cause and effect," the quantum factor "introduces a strong holistic element into our worldview." The decision, for instance, to treat subatomic entities as particles or waves "depends on the sort of experiment chosen" (Davies 1983, 107). Hence the two electrons in the EPR experiment must "be regarded as a single system" despite being "widely separated" (Davies 1983, 111). Moreover, it becomes "meaningless to talk about ... even the very notion of an atom, except within the context of a specified experimental arrangement."

Bohr hit upon the idea that dualities, such as subatomic entities manifesting both particle and wave behavior, despite being incompatible, can be combined in a complementary fashion. Correspondingly, the complementarity principle became the proposal of Bohr and his Copenhagen school. The experiment of John S. Bell provides a further corroboration of Bohr's outlook. "Whereas the hidden variables in classical physics are local mechanisms," intending by this to mean the contact of substances thereby allowing an interaction to occur, "those in quantum physics are nonlocal," or not dependent upon classical connections (Capra 1983, 343). In 1965, three decades subsequent to the EPR experiment, when Bell put it to the test, his "theorem dealt a shattering blow to Einstein's position by showing that the conception of reality as consisting of separate parts, joined by local connections, is incompatible with quantum theory." Thus Bohr's view which emphasizes the cosmos as "fundamentally interconnected, interdependent and inseparable," seems to be vindicated (Capra 1983, 346), at least provisionally.

Recalling a previous theme about relativistic continuity, determinism and predictability over against quantum discontinuity, indeterminism and unpredictability, the quantum model renders coherent those phenomena where instances of uncertainty and spontaneity are revealed and perhaps even abound. Mention was made earlier of radioactive material where "each individual nucleus will decay at an unpredictable time. This time will vary from one nucleus to another in a lawless way" (Bohm 1980, 67). Part of the difficulty rests with observer interference—the subject influencing the experimental results. Scientists and philosophers alike are driven to conclude that "[e]ither there will always be some hidden undiscoverable variable inherent in the act of observing, or there is a certain amount of unaccountable lawlessness at the bottom of things" (Owens 1983, 33).

A distinction must be made at this point between methodological and metaphysical principles of sufficient reason. Science deals with the former in that reasons are sought for what is observed or sensed. Yet in some instances science also appears to encroach on and dabble in the latter when it insists that such reasons *must* exist. There is a gap between descriptions of natural operations and prescriptions of how they ought to run. An illegitimate move has been made from the way things are to statements about the way they must be. Despite the confidence placed in the performance of science, the metaphysical approach lies outside the domain of conventional science. If science wishes to incorporate the second approach, then its borders will need to be expanded. There is nothing wrong with having an enlarged vision, but then a different agenda will need to be drawn up. Our central objective in this study is to determine if there are any promising proposals to this end.

The Anthropic Principle

The final stop on this tour will investigate how a metaphysical distinctive, namely *telos*, has been made the very hallmark of one camp in recent scientific speculation. The anthropic principle takes on a teleological flavor as a way of explaining how biology and biologists can arise. The main issue involves the physical constants and the way in which they are constructed, allowing humans to develop and comment on them. These constants display "delicate fine tuning" and a sensitivity whereby "[o]nly a minute shift in the strengths of the forces brings about a drastic change in the structure" (Davies 1983, 188-9). Some of the constants of nature include "the gravitation constant, the velocity of light, the electron and proton masses, [and] the electron charge" (Barrow & Tipler 1988, 225). John Leslie describes the coincidental nature of these constants in this manner:

Our universe does seem remarkably tuned to life's needs. Small changes in the strengths of its main forces, in the masses of its particles, ... in its early expansion speed, and so forth would seemingly have rendered it hostile to living beings of any plausible kind. (Leslie 1989, 25)

There is a tendency on the part of those who notice this fine tuning to inject metaphysical statements into the proceedings whether they are called for or not. Here, they might argue, is a place where philosophical and even religious notions can be reintroduced into an enterprise where they were evaluated as unwarranted before. They are driven to these conclusions from science, which gives them support for these views. Some even attribute the coincidences of the values of these constants to divine activity: "It is hard to resist the impression that the present structure of the universe, apparently so sensitive to minor alterations in the numbers, has been rather carefully thought out" (Davies 1983, 189). The cumulative weight of these coincidences point to the "deep enigma" that "[w]e who are children of the universe—animated stardust—can nevertheless reflect on the nature of that same universe" (Davies 1992, 232). But not all researchers are prompted to reflect on this sensitivity in the same way, particularly when supporting religious positions. Nevertheless, some do declare that here is one philosophical and scientific way in which a divinity can be reintroduced into the calculations in a respectable way, since it is reinforced by current scientific investigation.

Now back to our theme. The materials necessary for life-forms to thrive, at least those of which we are familiar, such as carbon and the heavier elements like iron, cannot be manufactured by the earth itself. Iron is produced by the fusion of helium, which occurs only in the interiors of large stars. Our sun is not large enough to accomplish this task. For there to exist iron on earth, so runs the current scientific reasoning, it must have been transported here from an exploding star, usually a supernova. Biological forms on earth, being carbon-based and, for those which bear it, having an iron atom in their hemoglobin, are thus fallout from stars. And with the onset of conscious entities, "the universe has generated self-awareness," urging some to believe that this eventuality "can be no trivial detail, no minor by-product of mindless, purposeless forces. We are truly meant to be here." Yet once again this can be regarded as debatable.

In any case, the first of two versions of the anthropic principle we will discuss here is the weak form, or WAP. It states that given the length of time that the cosmos has been in the business of world-building, there have come to exist "sites where carbon-based life can evolve" (Barrow & Tipler 1988, 16). This view borders on the tautologous, since "[i]t expresses only the fact that those properties of the Universe we are able to discern are self-selected by the fact that they must be consistent with our own evolution and present existence." An easy claim to make in retrospect. In other words, we see things the way we do because that is the way we have been constructed. "[A]ny observation we make is necessarily self-selected by this absolutely fundamental fact" that "we are a carbon-based intelligent life form" (Barrow & Tipler 1988, 3). Correspondingly, scientists are part of the very nature they study. One implication to draw from this definition is that scientific instrumentation is useful only to the limits set for it by the carbon-based intelligent life-forms employing them (Barrow & Tipler 1988, 23).

The second is the strong version, or SAP. In this rendition, "the Universe *must* be such 'as to admit the creation of observers within it at some stage'" (Barrow & Tipler 1988, 6). This second form goes well beyond the first and suggests that the cosmos "could not have been structured differently." Barrow and Tipler object, however, that we are privy only to a single universe at present, meaning "where do we find the other possible universes against which to compare our own in order to decide how fortunate it is that all these remarkable coincidences that are necessary for our own evolution actually exist?" Regardless of how unlikely the present status of the universe might be, ultimately it "had no choice about appearing with the appropriate degree of order required for life to appear" (Davies 1983, 171).

In the weak form one can infer prior circumstances from later consequences, such as humans exist, therefore carbon needed to be delivered to the earth in some way. But the principle explains neither the cause nor the effect, making the WAP minimally controversial (Drees 1990, 119). The strong version, on the contrary, yields a more metaphysical posture such that "any possible universe must have the properties for life or consciousness." Sadly, science by itself is not equipped to address these issues. Besides, some researchers maintain that confidence in the anthropic approach is waning since the coincidental nature of the physical constants is "eroding." This is because some instances "have found ordinary scientific explanations ... [while others] may in future be explained by theories of quantum gravity."

How It All Began: The Early Years

There is one more task to perform before we launch out into new ground, or set sail onto new waters, and this deals with cosmic beginnings. If one were to extend Einstein's general theory of relativity in the reverse direction one would theoretically approach the onset of space and time themselves. Prior to this point one cannot comment on the nature of time and space as understood by science. We are purposely avoiding the topic of universal origins since this is an area where science can only remain silent and our intent is to include it in the discussions. General relativity suggests that the expansion of the universe can be rewound, at least in our calculations, and a big bang model with an explosive cosmic birth can be formulated. An account of the early universe is presented by Willem Drees and is recounted below.

At the hypothetical point t=0, or time 0, all objects were in exceedingly close proximity, which is referred to as the point singularity. At this instance of infinite gravitation and density the undefinable nature of space and time can only be unlocked, if at all, with a quantum theory of spacetime (Drees 1990, 223). Next, at the (Max) Planck limit of time, namely t=approximately "ten raised to the exponent minus forty-three" (or 10^{-43}) seconds, "spacetime can be treated with the well-known theory of general relativity, [although] theories about matter are not known, or at least not tested independent of cosmology" (Drees 1990, 224). This latter instance is at least anticipated by mathematical models. The Planck limits describe "a boundary for the possibility of normal physical explanation ... calculated from the constants of quantum physics and gravity" (Drees 1990, 223). Planck's constant is a proposed limit to the microcosmic scale of space and time. It conveys the idea that there is no physical thing of length shorter than approximately 10^{-33} centimetres (nuclear dimensions are of the order of 10^{-13} cm.). Hence all physical interactions must occur on a scale greater than this length. So too with time. The time required for any physical change to occur is calculated as no less than 10^{-43} seconds (nuclear time frames are of the order of 10^{-23} sec.) (Barrow & Tipler 1988, 292). Thus physical interactions actually have a duration or completion time and do not occur at an instant.

Then for events at approximately t=one billionth of a second "theories of matter and spacetime [are] known and tested independent of cosmology," that is they are capable of being investigated under laboratory conditions (Drees 1990, 224). Finally, for t=approximately 300,000 years "the Universe becomes transparent" due to "the decoupling of matter and radiation" in the form of photons (Drees 1990, 219). This enables the cosmos to become visible for the first time since the displacement of free photons is requisite for our visual capacity. From this moment begins the formation of stars and galaxies.

Despite the violent imagery of this process the factors entering into it are extremely delicate—minor alterations would have had a drastic effect. The cosmos "risked collapse within a fraction of a second, or becoming a universe of black holes, or a universe of matter much too dilute to form stars and planets, or even one composed of light rays alone" (Leslie 1989, 25). In any event the big bang model cannot take us back any farther than the Planck time; from here a theory of quantum gravity might eventually prove helpful in unravelling the mystery. And with the natural restrictions encountered by the world to space, time, matter, gravity and the other physical forces, there is no way of knowing if there was to have been a universe "previous" to ours which recollapsed into a big crunch and then in turn bounced back into a big bang, thereby yielding our present universe. Such universal oscillations are forever beyond our reach. An entirely different set of natural laws might have obtained in those circumstances, meaning our conceptions of space, time, matter, gravity, etc. may not apply to any other universe.

The point of this moderately lengthy excursion into some of the details of the new (although by now slightly dated) physics, is to highlight the metaphysical language in which physics currently trades. Nor is this comfort zone confined to the physical sciences, for the biological or natural sciences are also familiar with it. An extended discussion of evolutionary theory, however, will be postponed until other topics have been broached so that evolution may be seen in their light. Subsequent to that, we will consider whether this type of "extramundane referent" language would be better suited in a different systematic orientation than the conventional one; that is, whether it would find a more congenial hitching-post elsewhere, even in something other than a substance view of reality. But this is to anticipate an upcoming theme.

The Story Thus Far

It is somewhat cynically said that the eminence of a scientist is measured by the length of time that he [or she] holds up progress in his [or her] field. (Lovelock 1979, 70)

By way of summation, let us review the ground upon which we have trod from a somewhat different angle. To begin with, the presupposition commonly referred to as the universe's intelligibility and contingency is what enables empirical science to proceed (Davies 1992, 169). That the universe could have been otherwise is vital for the scientific enterprise; were it not contingent "we would in principle be able to explain the universe using logical deduction alone, without ever observing it." But a contingent universe "cannot contain within itself an explanation for itself" (Davies 1992, 171), nor can the basis of something be explained in terms of that same something. At the same time, that which appears contingent "might be shown by future science to be unavoidable, given certain circumstances" (Drees 1990, 99). Science then simultaneously relies on contingency and works toward necessity, making this endeavor "a quest to remove contingency as far as possible," so that indeterminism should be replaced by the law-abiding. As for intelligibility, nature is usually understood as rational, prompting the acceptance of the principle of sufficient reason (Davies 1992, 162), that there must be a logical explanation for what occurs. Yet one must ask for the basis of one's confidence in this approach: "Is there sufficient reason to believe the principle of sufficient reason?"

Building theories, as we have established, also contains as much in the way of aesthetic or non-rational factors as it does rigor (Davies 1992, 177). Moreover, the danger exists that the metaphors which are imported into research and which fuel scientific assumptions may also (inadvertently or otherwise) be read off the results: "when the research is later analyzed," what one may obtain "is more a retrieval of these presuppositions than a discovery of the intrinsic implications of the research itself" (Russell 1993, 29). Both religion and science employ metaphorical language, yet this need not be viewed negatively, since the practice is "unavoidable and essential for any culture, providing an expression of its self-understanding" (Drees 1990, 162).

Myths, though, are perceived as "outdated fables" and therefore either scientifically untrue or, what is worse, irrelevant and even meaningless to a mature society. Mythical language utilizes "multivalent symbols, whose referent is ... the transcendent or the sacred" (Gilkey 1981, 66). While the task of theology is the "reflection on the mythical symbols of a tradition," science has dismissed myths as useful only in the service of the so-called soft sciences (Drees 1990, 162). This mindset has promoted the drive to transform disciplines so that they will display a more scientific stamp. The flagship field for such imitation is physics and to accomplish the feat of bearing a resemblance to it, disciplines in the social sciences, arts and humanities have sometimes taken a more statistical approach to research in their areas. Yet two issues still need to be addressed. First, science has misunderstood the value and proper role of myths as pointers to the transcendent; and second, it has also failed to recognize that "as a form of human inquiry," science "points beyond itself to a ground of ultimacy which its own forms of discourse cannot fully thematize, and for which religious symbolization is alone adequate" (Gilkey 1981, 41).

Religion and science then appear to be more closely related than previously anticipated. Davies perceives science not as a quest for truth but about creating pictures or models of reality that allow "us to relate one observation to another in a systematic way" (Davies 1983, 219). This type of creativity overlaps with more artistic pursuits. What becomes clearer is that science can "tell us what we can *know* about the universe, not how it *is*." According to the instrumentalist version of knowledge, models and theories are neither right nor wrong but more or less useful in the way they organize and connect "a wide range of phenomena in a single descriptive scheme." Davies submits that "[b]y basing itself on utility rather than truth, science distinguishes itself sharply from religion" (Davies 1983, 220). Drees takes on a similar though somewhat moderated position in admitting that both science and religion are the two main routes for "exploring reality" and that they also "may share tools," but he then cautions that this "does not by itself diminish the separateness of their contents" (Drees 1990, 160).

The approach of Gerhart and Russell is perhaps the most nuanced of the ones we will examine. They advance the notion of the "metaphoric process" which enables "genuine new knowledge [to come] into existence. In that sense we create the world, as we understand it, by creating the understanding" (Drees 1990, 166). The two authors expand on this theme in the following way:

> 1. Science and religion are different fields of meaning, but have the same epistemological structure. The same process, analogical extension and metaphoric creation, are involved in the formation of new knowledge. And 2: Relating concepts from the scientific field with concepts from the religious field is also such a metaphoric process curving the existing worlds of meanings.

By being related they are each strengthened. Theology is needed, since it gives a theoretical status to experiences of limit and transcendence. A claim to truth is less probable for any understanding which does not relate the two fields.

What becomes increasingly plain is the limitation of our field of vision. We have trouble seeing clearly, indeed we cannot, and the reflections on our experience which both religion and science engage in are often aesthetic constructions grounded in personal considerations. Fundamental questions persist about "the mystery of why the universe has the nature it does" (Davies 1983, 43); why "things [are] the way they are," "[w]hy *this* universe, *this* set of laws, *this* arrangement of matter and energy? Indeed why anything at all?" (Davies 1983, 46).

I find the following analogy helpful when probing our perspectival limitations. We as humans boast the capacity to perceive our environment, but the extent or depth of our perception is what is at issue. Consider a pet such as a dog in a room together with its owner who is watching television. The dog is exposed to the same information from the television programming although it appropriates it differently, what with heightened sensitivity to sound yet with color blindness and a smaller neural network. The dog filters the information in an alternate fashion—it might be aware of sights and sounds emanating from the set but it lacks the requisite symbol system to work with the information in the same way as its human master. Relying heavily on smell, for example, the dog is unmoved by the set's lack of olfactory cues. Hence much of the material is lost on the pet.

The same may be true for us as we attempt to piece together the puzzles of nature and try to comprehend what is taking place, say, at the quantum level. We approach these phenomena with our own categories of what we understand, assume and are willing to accept. This may be similar to a dog whose master frequently goes away and leaves the dog at home only to return after a while. To the dog the world is comprised of masters who either mysteriously disappear and reappear or are wilfully hiding. We may resemble this pet in applying our categories to situations which perhaps call for a modification in our perceptual framework. As we currently perceive the subatomic world, for example, it seems at times to be random, even chaotic; but at another level it might be interpreted as orderly. The conceptual gap between the dog's and its master's cognitive apparatus does not allow the dog to benefit from the television experience in the same way as the master, to say nothing about whether it even boasts the same experience or whether it has the same taste in programming. The same may occur with scientists who come across items which do not fit their categories or conform to their "pet" presuppositions. Either our senses have not been sufficiently trained so as to perceive the new phenomena adequately, or like the dog much of the experience is lost on us owing to a deficiency in our capabilities. In the capacity of making sense of our experiences there may be commonality here with "canine logic."

In view of the world of quantum mechanics, indeterminism has carried the day over determinism and therefore Bohr over Einstein. Contrary to Einstein, not only does God appear to play dice but God actually seems to relish the occupation. The quantum world of a dice-playing God could even be regarded as germane to divine activity, yet this is a matter of perspective since "[w]e either see the evidence of God everywhere, or nowhere" (Davies 1983, 70). It is futile to expect God to reside in the gaps of our knowledge because God often gets squeezed out of those spaces. Instead, "[i]f God is to be found, it must surely be through what we discover about the world, not what we fail to discover" (Davies 1983, 209).

Scientist-theologian Arthur Peacocke takes any true "inquiry into the nature of reality" as a legitimate pursuit, whether scientific, philosophical or otherwise, in that it abides by "the same general criteria of reasonableness" and seeks "the best explanation" that fits the data (Peacocke 1993, 91). Science by itself, though, he contends, is incomplete because it can tell us nothing about our own subjectivity (Peacocke 1993, 110). Peacocke suggests that one way of imagining the interaction of God with the world is through the mind-brain model (Peacocke 1993, 161). Basing an analogy on something about which we know very little, however, is not fruitful, since it is precisely the workings of the mind-body relation that are baffling. Were more of it known, we could then assess more accurately whether God interacts with the world in a similar fashion.

The interim conclusion to be drawn is that with respect to the glimpses, hints or traces of divinity in the cosmos, it depends on what one is looking for and open to. Not only do metaphysical elements and religious convictions play a role in theory development but they incline one to subscribe to certain scientific concepts over others (Drees 1990, 67). For instance, if a scientist entertains "a strong theological interest in history," this person "would prefer a metaphysic which incorporates time asymmetry" (Drees 1990, 67-8), where there is a direction to events or an arrow of time. Theological commitments involving God as active in historical events would opt for this strategy. Alternately, if God is construed as "ultimate, atemporal transcendent Ground," then one would likely select a research program more in accord with this outlook—specifically one which de-emphasizes history or time. But we have ventured as far as we can within these parameters. What lies ahead is a refocusing of standard issue interpretive lenses so that conventional categories may be recast. We are now equipped to open up a new map and embark on a new journey.

CHAPTER THREE

The Reenchantment of Nature

Indeed, very little of what could be called direct sense perception takes place in physics today ... the connection between experimental apparatus and human experience is becoming increasingly remote. (Bohm & Peat 1987, 45)

The scientist, even when he is a believer, is bound to try as far as possible to reduce miracles to regularities; the believer, even when he is a scientist, discovers miracle in the most familiar things. (Hooykaas 1969, 206)

The subjective experience of wonder is a message to the rational mind that the object of wonder is being perceived and understood in ways other than the rational. (Zukav 1980, 40)

THE NEW JOURNEY is actually an old journey reconsidered. Here is what I mean. There is another way of writing a history about the relation of religion and science than the ones provided by Draper and White. Morris Berman directs his attention to the enchanted view of the universe during the time previous to the scientific revolution. The tale he recounts on this score involves nature as alive and the inhabitants of the world as important ingredients in the unfolding of its drama (Berman 1981, 16). This fostered a sense of belonging which Berman refers to as "participating consciousness." With the rise of the scientific worldview, however, the account becomes "one of progressive disenchantment." As scientific explanation came to deal with matter in motion and "insist[ed] on a rigid distinction between observer and observed," this evoked a non-participating consciousness on the part of the world's inhabitants and was followed by a sense of alienation from it. Here pure observation was thought to be possible through the detachment of subject from object.

In Newton's mechanical philosophy, drawing on the Cartesian model, the knowledge of an object meant "to subdivide it, quantify it, and recombine it" (Berman 1981, 45). Armed with this new approach, a shift occurred on the part of these practitioners of natural philosophy "from quality to quantity, from 'why' to 'how.'" Aristotle's use of four causes—material, efficient, formal and final—which to that point had been adopted in the description of events, became reduced to two as the latter two outlasted their usefulness. Nature, far from "possessing its own [plan,] goals and purposes, [became] a collection of inert matter." In the medieval world picture, physics and metaphysics, epistemology and ethics are still combined: "'What do I know?' and 'How should I live?' are in fact the same question" (Berman 1981, 51). In the modern period, though, there is no "immanent meaning" but only atomistic descriptions of nature. To the people of the middle ages, "[e]verything but God is in the process of Becoming" (Berman 1981, 50), and their understanding revolved around the conception that "all material events and processes had psychic equivalents and representations" (Berman 1981, 92). Our "premodern counterpart[s]" pictured themselves as identical to their experiences (Berman 1981, 77). In the modern world, on the contrary, "knowledge is obtained by distancing oneself from the experience" (Berman 1981, 139).

It is a commonly held view that the "official scientific culture" enjoys "an absolute, transcultural truth" (Berman 1981, 50). Berman endeavors to undo this line of thinking. In the list of thinkers who began to notice cracks in science's epistemological edifice, Berman cites Kant who "was perhaps the first Western philosopher in the modern period to recognize that the mind is not simply bombarded by sense impressions, but actually plays a role in shaping what it perceives" (Berman 1981, 317, n.1 (c)). Berman also mentions the work of Polanyi who sees rationality as functioning viscerally before it operates cognitively (Berman 1981, 139). By "attributing truth to any methodology we make a nonrational commitment," which in Polanyi's estimation amounts to "perform[ing] an act of faith" (Berman 1981, 136). In as much as "the knower is implicated in the known" (Berman 1981, 139), this reinstates Berman's category of participatory consciousness. What was thought to be long buried by virtue of the rise of science has now resurfaced through the realization that "the knowing person" participates "in the act of knowing." Observation on the part of subjects interferes with objects and ultimately affects both.

Berman reinforces the points we made earlier concerning positivism which holds that one need only "[a]bstract a thing from its context ... and the situation becomes meaningless; although perhaps mathematically precise" (Berman 1981, 248). Yet crediting Heisenberg, a contemporary of this positivistic effort, our observations are now understood as not purely about nature "but nature exposed to our method of questioning" (Berman 1981, 145). Accordingly, what we are left with is not confidence in scientific objectivity but the realization "that there is no fixed reality, ... only ... knowledge adequate to the circumstances that generated it" (Berman 1981, 150). Curiously, after about a century of quantum theory, its principles have failed to penetrate the popular view that nature's knowability is independent of human factors. Science in the main continues to operate without the awareness that observation "alters the knowledge obtained" (Berman 1981, 184).

Science has become "the integrating mythology of industrial society;" though in view of its epistemological difficulties "the whole system is now dysfunctional" not yet four "centuries after its implementation" (Berman 1981, 193). Berman presents the work of philosopher Susanne Langer who deftly sums up the current situation. In referring to her landmark volume *Philosophy in a New Key*,[7] Berman announces that

56   *God and the New Metaphysics*

> A new key in philosophy does not solve the old questions; it *rejects* them. The generative ideas of the seventeenth century, she says, notably the subject/object dichotomy of Descartes, have served their term, and their paradoxes now clog our thinking. "If we would have new knowledge," she concludes, "we must [invest in] a whole new world of questions." (Berman 1981, 183)

For Berman the world of new questions is located in the work of Gregory Bateson. While Berman acknowledges the contributions of Whitehead he elects instead to concentrate on Bateson. Over against the modern conception of science Berman lists Bateson's counterproposals of a reenchanted science:
1) where fact and value are inseparable;
2) where our relations with natural phenomena find their proper context;
3) where measurement of the concrete is fitted alongside the abstract;
4) where mind and brain/body together with subject/object are double "aspects of the same process;"
5) where infinite progress is rejected;
6) where "both/and" logic takes precedence over "either/or";
7) where contrary to reductionism wholes contain "properties that parts do not;"
8) where mind is "not reducible to [its] components;"
9) and where nature is not seen as dead but alive, that is, reanimated (Berman 1981, 238).

Now let us see how Whitehead fares in comparison.

Process as Reality

> *[M]atter has potentialities unrecognized and unheeded by a reductionist materialism.* (Peacocke 1993, 245)

> *[The right hemisphere of our brains where intuitions are said to occur] has not atrophied from lack of use, but our skill in listening to it has been dulled by three centuries of neglect.* (Zukav 1980, 40)

The Reenchantment of Nature 57

Alfred North Whitehead (1861-1947) endorsed what Berman refers to as participating consciousness or the reenchanted view. Two-thirds of a century after the publication of Charles Darwin's *Origin of Species*, the division between science and religion was addressed by Whitehead in his Lowell lectures of 1925, later published under the title *Science and the Modern World*. It was here that he sought to repair the antagonism between the two fields and show that they could be reunited. In his main corpus of philosophical writings he goes about this task in the following way.[8]

As early as 1933, Whitehead anticipated certain views of authors such as Popper, Kuhn and Polanyi by declaring that "reports upon facts are shot through and through with theoretical interpretation" (Whitehead 1967a, 3). Interpretation unavoidably enters scientific undertakings, for "[k]nowledge is always accompanied with accessories of emotion and purpose" (Whitehead 1967a, 4). Further, he boldly asserts that "observation is selection" (Whitehead 1967b, 18). Any type of induction, scientific or otherwise, for Whitehead "presupposes metaphysics" (Whitehead 1967b, 44). This has implications for historical disciplines since there is no "rational justification for [one's] appeal to history till [one's] metaphysics has assured [one] that there *is* a history to appeal to; and likewise ... that there *is* a future already subjected to some determinations." If positivistic science deals only with observation, then it may not comment on the future since it cannot be observed (Whitehead 1967a, 125). Conjectures "beyond direct observation spells some trust in metaphysics," regardless of how unwittingly these ideas might be held (Whitehead 1967a, 128). For Whitehead metaphysics serves to "guid[e] imagination and justify purpose," and is not something which science should fear or seek to eradicate. In fact it is precisely the "urge towards explanatory description [that] provides the interplay between science and metaphysics" (Whitehead 1967a, 128-9). What is more, the pursuit of rational "justification of brute experience" has supplied the incentive for scientific progress, and in this way science betrays "a variant form of religious interest" (Griffin & Sherburne 1978, 16).

Whitehead (167a, 309) refers to his counter-cosmology as *organic realism*, the *philosophy of organism* or *organic mechanism*

(Whitehead 1967b, 80). He gives priority not to being but becoming, and in accord with his magnum opus *Process and Reality* he submits that process is the reality (Whitehead 1967b, 72). Albeit counterintuitive, everything else is an abstraction, and to mistake "the abstract for the concrete" he diagnoses as committing the fallacy of misplaced concreteness (Whitehead 1967b, 51). The notion of the universe as composed of "static stuff" is substituted with "fluent energy" where everything becomes a type of energy in flux (Griffin & Sherburne 1978, 309). There are no substances which exist independently as in materialism (Whitehead 1967b, 152). Rather, as Whitehead postulates, "all final individual actualities have the metaphysical character of occasions of experience" (Whitehead 1967a, 221), which become "evolving processes" that comprise the natural world (Whitehead 1967b, 72). These entities are "spatio-temporal unities" or "events," which are "the unit[s] of things real" (Whitehead 1967b, 152). The concept of community is important in this scheme since "each relationship enters into the essence of the event; so that, apart from that relationship, the event would not be itself" (Whitehead 1967b, 123). This social view of reality stresses internal relations as remedial to the scientific overemphasis on spatio-temporal relationships as being fully external.

Whitehead's contention is that the undertakings of naturalistic science apply only "to very abstract entities" as a type of limit case, whereas "concrete enduring entities are organisms" (Whitehead 1967b, 79). Here "the plan of the *whole* influences the very characters of the various subordinate organisms which enter into it." An electron for example "within a living body is different from an electron outside it, by reason of the plan of the body" which "includes the mental state." These organisms, which can be as minuscule as subatomic particles, "differ in their intrinsic characters according to the general organic plans of the situations in which they find themselves" (Whitehead 1967b, 80).

Whitehead parts company with conventional positivistic science in favor of a universe displaying a dual nature without becoming extended into a full-fledged dualism. For him the universe "is dual because each final actuality is both physical and mental" (Whitehead 1967a, 190). As opposed to the dualism of the traditional Cartesian variety, in which there is a dichotomy of mind and matter, Whitehead combines them into a dipolar character of all entities, God included. In his cosmology both God and the world are immanent in as well as transcendent to each other, and each also creates the other (Griffin & Sherburne 1978, 348). In this way, "God is not to be treated as an exception to all metaphysical principles, invoked to save their collapse. [Instead, God] is their chief exemplification" (Griffin & Sherburne 1978, 343). What this amounts to is that the distinction between God and the world is largely, though not exclusively, a quantitative and not a qualitative one. In essence, "each temporal occasion embodies God, and is embodied by God" (Griffin & Sherburne 1978, 348).

Whitehead describes God as "that actuality in the world, in virtue of which there is physical 'law'" (Griffin & Sherburne 1978, 283). Furthermore, God's purpose according to Whitehead is the maximization "of value in the temporal world" (Whitehead 1974, 97). From the above reasoning we can regain hope for the religion-science relation because in Whitehead's reassessment they interact at each step. Mind and matter are rejoined, and physics and metaphysics are reunited. This means that science and religion are not fundamentally at odds and the universe can once again be viewed as enchanted.[9]

David Ray Griffin, himself a Whiteheadian process thinker and former Director of the Center for Process Studies, elaborates on this reenchantment theme. Griffin bemoans the positivistic accounts of science where "all downward causation from personal causes and all action at a distance" are ruled out (Griffin 1988a, 4). The epistemological reductionism of science regarding the capacity to know only that which is "understandable in materialistic terms" (Griffin 1988a, 5) promoted the ontological reductionism of "all causation [as] run[ning] sideways and upward, from parts to parts and from parts to the whole" (Griffin 1988a, 15). That wholes enjoy properties which parts do not is of course evident in emergent cases such as the wetness of water, where that which obtains at the macro-scale fails to occur at the level of individual constituents. Nor does one need to be a process thinker to appreciate this point. But like other critics of positivism, Griffin also takes notice of personal involvement in the scientific enterprise, and he renders the judgment that in place of "standing as an impartial tribunal of truth, ... science is seen as one more interested participant" (Griffin 1988a, 8). Griffin therefore rejects the epistemological imperialism of science—that is, the extent to which it aims at "providing the only genuine" knowledge of the world (Griffin 1988a, 6).

The disenchanted view of nature involves *inter alia* a division between explanation and understanding, which sets in motion a practice whose trajectory leads ironically to a disenchantment of science itself (Griffin 1988a, 3). Griffin captures the mood in this way:

> If all human life is meaningless, then science, as one of its activities, must share in this meaninglessness. For some time, many held that science at least gives us the truth, even if a bleak one. Much recent thought, however, has concluded that science does not even give us that. The disenchantment is complete.

According to the scientific world picture, matter is all there is, thereby rendering suspect the category of mind. Materialism seeks to avoid dualism yet fails to account for "the experience of experience itself—without covertly introducing dualism" (Griffin 1988b, 147). When faced with difficulties such as the mind-body problem, scientists tend to consider the "communication across the [ontological] gap" to be inexplicable (Griffin 1988a, 18). In other words, if matter and mind are qualitatively different then "their relation is unintelligible" (Griffin 1988b, 147). Better in their view then to dispense with the category altogether. But Whitehead's doctrine, which Griffin calls *organicism* or *deep empiricism*, treats all entities as bearing traits ordinarily attributed to mentality. In addition, they are not fully determined but, consistent with the quantum outlook, retain an amount of spontaneity or "self-determining power" (Griffin 1988b, 150) which renders the future authentically indeterminate.

The organicist perspective on reality "holds that all primary individuals ... exercise at least some iota of purposive causation" (Griffin 1988a, 22). Primary organisms are organized along two lines: they can be manifested either "as a compound individual, in which an all-inclusive subject emerges," or "as a non-individuated object," having no "unifying subjectivity." Members of the first class include animals; the second includes sand, rocks and trees. This implies that "there is no ontological dualism" in nature, only "an organizational duality." The fundamental units of reality are termed *actual entities* and they do not endure, instead they are momentary events which perish upon the completion of each process of concretization (Griffin 1988a, 23). For each such round of process, "[e]fficient causation still applies to the exterior of an individual and final causation to the interior." In Griffin's appraisal, "because an enduring individual, such as a proton, neuron, or human psyche, is a temporal *society* of momentary events, exterior and interior oscillate and feed into each other rather than [run] parallel."

Yet as intimated, not all individuals endure. Those occupying what is customarily referred to as the inorganic realm lack experience as wholes, despite their constituent individuals enjoying it by virtue of being individuals, and hence are named *aggregates*. Genuine individuals in contrast "have (or are) experiences" (Griffin 1988b, 152). At bottom "all actual things either *are* experiences or are *composed* of individuals that are experiences." This means that there is nothing foundationally inorganic in the world since all entities display organic qualities. It is for this reason that Griffin refers to this scheme as *psychicalism, panpsychism,* or *panexperientialism* where everything is experience. This is no substance view of reality but an experiential understanding of the "stuff" of the universe. Since mind and consciousness for Whitehead reside only in the higher organisms and since sentience is fundamental, Griffin prefers the last of these three terms (Barbour 1990, 225).

In this Whiteheadian strategy, "the most primitive form of perception" is pre-sensory (Griffin 1988b, 153). Each occasion of experience "combines receptivity (physicality) and self-determination (mentality)" (Griffin 1988b, 154) as it organizes, synthesizes and integrates its experiences. At the completion of each round of becoming, the experiencing subject becomes an object for the next developing occasion such that "a subject for itself is now an object for others." Most importantly, in Griffin's estimation, "*[s]ubjects and objects are not different in kind, merely different in time.*" (Griffin 1988b, 155). To reiterate a previous point, certain particles at the atomic level are enduring things in their wholeness, but not necessarily so with the individuals of which they are composed. Protons endure, yet the quarks of which they are allegedly made might not. Each enduring thing is a society of individuals, known as "a series of rapidly occurring events." The lifespan of an enduring thing such as a human psyche may be eighty years or so, but a specific molecule within it may have a duration of several billion years (Griffin 1988b, 157). On the other end of the scale, some subatomic particles might exist on the order of nanoseconds and may experience just one or a few rounds of becoming. Their lives would then not be very "event"ful.

There are many other respondents to Whitehead's cosmology, some of whom will be mentioned below. In each case it can be identified that Whitehead's organicistic metaphysics provides a groundwork for Berman's reenchantment project. James E. Kirk, for one, notes that Whitehead's system was given an indifferent reception, owing largely to "the rift between science and philosophy in this century" (Kirk 1991, 79). The Newtonian model has been superseded by making the quantum and relativistic worlds central and the world of Newton peripheral. Yet this does not imply that science is any more willing to embrace or even entertain Whitehead's ideas. If methodological science is translated into worldview science then it will certainly influence our perception of the cosmos, but as Willis W. Harman urges, "*the way the world is experienced in our culture* [also] *influences what kind of science gets developed*" (Harman 1988, 121). Precision, simplicity and objectivity are unattainable goals, meaning that an idealized version of the universe containing "point masses, frictionless surfaces, instantaneous velocities" and so forth are, as Frederick Ferre contends, "not directly met in the unkempt world of experience" (Ferre 1988, 88). The situation science faces is that nature is fraught with the irregular. The fixed and rigid are often imported notions, whereas Whitehead upholds spontaneity and makes creativity the ultimate basis upon which he builds his view of reality.

Science is left to trust that the poorly fitting will eventually conform to accepted practice. Whitehead, on the contrary, is not threatened by spontaneity but rather welcomes it. It enjoys a place in his science because it has a place in his ontology, since his fundamental units of existence each display some measure of it.

Science in its quest for objectivity seeks to avoid reference to subjectivity, meaning that "our own experience is deemed irrelevant" to legitimate research programs (Kirk 1991, 56). Whitehead, though, makes knowledge depend on it. As mentioned, this is not a substance view of reality in which the present world is comprised of particulate objects, but an experiential framework where developing subjects enjoy interiority. Relations are not restricted to external ones, suggesting that, for example, hydrogen atoms on the surface of the sun are different from hydrogen atoms on the surface of our skin by virtue of their internal relatedness. Such is Whitehead's interconnected understanding of reality wherein, as John B. Cobb, Jr. takes care to remind us, "things cannot be abstracted from relations to other things" (Griffin 1988, 107). Modify the "set of relations," and "the things themselves" are likewise altered.

Entities are thereby "constituted by their relations" (Griffin 1988, 108), making 'structure' a "fundamental category of analysis" (Griffin 1988, 107). That arrangements of entities and their internal relatedness should produce properties where they did not occur in any degree before, hence resulting in a difference in kind, is most assuredly a "nonmechanistic notion." A substance view does not account for emergent properties; and if in the materialistic claim that only matter endures, then "in what sense can an organism which replaces every cell of its body over a period of time be said to 'endure'?" (Kirk 1991, 59). In contradistinction, the process approach takes "[t]hat which endures [a]s that which evolves: a pattern of activity which is constituted by its ongoing reevaluation of its relationship with its environment. It is the 'pattern' which marks out the organism as an enduring 'thing.'" Whereas matter is rearranged, experience is acted upon. Each occasion can creatively incorporate its own immediate (and even more distant) past, together with data from other entities in its immediate (and even more distant) past, into the presently developing entity. And in so far as it can elect to synthesize and integrate these data in genuinely new ways, novelty arises in the world. "This is the reason," Kirk concludes, that "entropy cannot be the last word in an organic universe" (Kirk 1991, 219).

Commenting on emergent properties, Davies goes so far as to state that science might have reversed the order in assuming that mind arose as the organization and complexity of matter increased (Davies 1983, 71). Perhaps intelligence arrived first on the scene, implying that mentality could be the primordial category (Davies 1983, 210). Whitehead differs from conventional science in that he believes the latter manufactures a truncated view of reality where what matters is matter. Instead, material reality for Whitehead is not the concrete but the abstraction and what matters for him is internal relations and the responses of entities to them.

One entity to which occasions are related is God who provides for them an ideal which they might wish to grasp. God presents them with an initial aim, like truth, beauty, goodness and the like, and occasions compare it to their own subjective aims. Should there be a conformity of aims, then the occasion applies it to its own integration process. God's sphere of influence is therefore universal. Some authors such as Davies and Drees acknowledge the Whiteheadian scheme but stop short of evaluating it as a serious contender to mainstream science or more traditional forms of Christianity. John Polkinghorne admits that while such a view has its appeal, he finds difficulty in describing entities by using the category "chains of events" since this makes reality strangely "episodic" for him (Polkinghorne 1989, 14). His dissatisfaction with Whitehead's notion of enduring individuals which contain entities that perish lies in this exceedingly discontinuous conception of nature. Whitehead's approach does, however, seem to cohere nicely with the quantum understanding of the world. Both Polkinghorne and Peacocke, however, are prepared to accept the emergence of consciousness from a specific arrangement of materials. Whitehead would agree that arrangement or structure is crucial for such properties to emerge, as they are fuelled by the relations between the components, yet would add that materials are not the issue here but the presence of a social order boasting an occasion with an organizing center which is sufficient to propel this capacity.

Authors such as Drees, Peacocke and Polkinghorne admittedly draw valuable insights from Whitehead's ideas, but their objections persist. For instance, Polkinghorne is suspicious that Whitehead's panentheism as his followers have termed it—the view that God contains the world and evolves along with it rather than is identical to it as in pantheism—will upset "the mutually free relation of God" and the world if they are to be taken as coextensive (Polkinghorne 1989, 22). Yet at the same time he supports the Whiteheadian category of dipolarity—the idea that each entity enjoys and exercises both physicality and mentality. He imports some of Whitehead's insights into his own scheme, but the strategy of incorporating process themes into traditional frameworks is potentially problematic. The operation may not be a straightforward success especially if such a grafting technique is rejected by the host paradigm.

As can be appreciated, Whitehead is not without his critics; but we may tentatively conclude that at least he provides one system that reunites science and religion through the dipolarity of reality and the God-world relation. Physics is reconnected to metaphysics and God incorporates them both. The advancements of contemporary science appear to resonate with the process outlook. This could provide science with a new paradigm that consolidates experience and intentionality throughout the entire scale from quantum to cosmos. The Aristotelian formal and final causes would then be reinstated alongside material and efficient ones. The proposal that Whitehead offers might be a step in this direction.

A Sporting Chance

Analogies assist us in rendering unfamiliar concepts into familiar terms. In an attempt to get a handle on what is occurring at the Whiteheadian metaphysical level, it seems appropriate to draw an analogy from the world of sports, namely the North American pastime known as football. For me at least, this sport most facilitates the notions that Whitehead is trying to convey. Football is a team sport that has definite plays which exhibit specific start and stop points. This is typical of football which operates with a series of "downs," but it is also true of baseball where the action is subdivided into a certain amount of "at bats" each inning for the offensive side and within that a number of pitches for the defensive side. I elected to go with football since baseball will feature prominently in an upcoming segment.

Whitehead's metaphysic involves the activity of experiences which occur and combine to form an event. The world is made up of occasions of experience which develop, become fully formed and then pass away into the past. They have their day, or more precisely their brief moment of occurring and they then take their place in history. The potential becomes the actual and then the historical, and always in the sequence of becoming and being-perishing. As objects they potentially contribute to new subjects.

Memory is a faculty which humans enjoy that can recall past events. What after all is one more intimately privy to than one's own experiences and memories of them? We can retrieve not only the factual data or the details of these events but also how we felt about experiencing them at the time. All of this is useful information for what we might like to do with it on a subsequent occasion. If it is pertinent to our new circumstance, then we will apply it. This means we will draw upon what we have at our disposal in order to respond to a current situation. Our relations to the data in our past world will influence how we feel about them, which in turn will inform how we respond to them. Our new circumstance may prompt us to respond in new ways. Such a moment of creativity means that something novel has entered the world which was not there before. And hopefully the world is richer for it. All of this is background so that the football analogy will begin to make sense.

The past impinges upon the present with what has gone on before, and current events draw upon it. From within the confines of a single football game, the kickoff has no past to speak of (other than previous football matches) except perhaps for the coin toss which decided which team would receive the ball. In the first series of downs, however, there is already a past to draw upon. Specifically, the kickoff return would have given the receiving team its initial field position. After the first play from scrimmage, there are additional aspects of the past at the disposal of the offensive side. The second down is what that team has inherited from its past exploits. The teams can do nothing to change it once the officials have spotted the ball and whistled in the next play. All that the players can do is respond to it. They are related to the past in that it displays to them their present position. Now the past is prompting them to act again based on the significance of the past for them. If the first play from scrimmage was a success for the offensive side, then they will look favorably upon their past and likely take that same attitude to the next play. The same obtains for the defensive side, though in the reverse way.

Since this is a team sport, the players need to coordinate their efforts in order to keep the drive alive. If the defense has held the offense to a meager gain on first down, then this is one of the signals the offense receives from the past. They then use the import of this message, marshall their strength and forge ahead with another play. Each play from scrimmage is analogous to an event in Whitehead's process scheme.

On the subject of coordinating their efforts, each team on the field is a society which has an organizing center. It is the responsibility of the quarterback to use the past to the team's advantage. Each team member has a role to play, but the quarterback takes all the necessary information from the past, an understanding of the athletic capabilities of his team members and makes their energies conform to the overall objective of the team, namely scoring points each time they enjoy ball possession. This is analogous to the subjective aim on the part of Whiteheadian occasions of experience.

Each player has an assignment and it is not carried out in a haphazard fashion. If this were the case, perhaps nothing would be accomplished—all the individual efforts, if not coordinated, might cancel each other out. But as it stands, the quarterback engineers the drive and plots a course for the team to follow. Since he has such a pivotal part to play, he could be considered the dominant member of each event. He is what makes this society more than simply a collection of individuals, or an aggregate, but a compound individual—one that through his navigation makes them work as a single unit. He unifies their energies for a common purpose—the common good of advancing the ball.

But as was mentioned previously, there is another aim at work in the process strategy. God is said to present to each entity, as it is about to get under way in its round of development, a place to develop and a goal to aim toward. The entity then decides the extent to which God's aim conforms to its own. In our football analogy the coach or the offensive coordinator, particularly if they have a say in what plays are to be called by the quarterback, assume the role of God (a piece of information which one might not wish to divulge to these persons lest they suffer ego-inflation).

Going into more detail, the play which the quarterback will decide to call depends on a number of factors including additional items from the past. These include the practice sessions that the team has held in preparation for this opponent. The events leading up to the game include those who displayed their talents in practice and so are deemed ready for game action and have been cleared to suit up for the game. There are also those on the injured list as well as the reserves or replacements on the bench in case of injury or ineffective play. This dictates the resources which the coaching staff has available and the attack which the quarterback can direct. The latter also has a knowledge of what plays have worked in practice and are likely to have success in the game. On the field on game day, s/he takes all of this into account as s/he plans the next move. This is analogous to how nascent, that is newly-forming, entities draw upon and respond to their past and how the past impinges upon the present, at least the way Whitehead sees it.

Now the quarterback has no guarantee of what play in his or her arsenal will have the most success, but s/he is free to judge. The information of the past is evaluated according to how well it conforms to the goal of ball advancement. The common good for the team includes the proper execution of plays and this is presented to them by the coaches. To this end, some data from the past will be germane to the new situation and others not. With the limited number of options that s/he has at his or her disposal, the team approaches the ball and another play commences. The play continues until the action is whistled dead by the officials. This point marks the completion of one event or round of becoming in process terms. The use of the term "dead" is appropriate here because in the Whiteheadian sense the moment has perished. It now assumes the role of a datum for the decisions to be made about the next play.

While the play is still under way the athletes on the field influence the outcome. Their subjectivity can work to alter the course of the event. Yet once the play is over their interests and efforts no longer affect the play that was. Their input will need to await the following play. As the completed play stands it is an objective datum for the statisticians.

Each new play is based on the set of all previous plays and so is a unification of them. What the previous plays have taught the team goes into the selection of the subsequent play. In this way "the many become one and are increased by one" as Whitehead would say. Not only can one draw upon the experiences of previous plays but the past makes itself felt for the upcoming play. If adjustments need to be made based on a lack of success with the previous play, then that will have an impact upon whether the quarterback will reevaluate the skills of the players and perhaps even reconsider the alternatives in the play book. With each successive play there is one more event in the world to draw upon. What the team's activity has bequeathed to the world is one more play, together with all the experiential information it bestows. All the many factors entering into that play are united in it and result in one more item to reckon with. The outcome now becomes one more factor in how the team will approach the next play. The team's internal relatedness to the previous play will definitely have a bearing on what follows.

Carrying the analogy even further, if a play is attempted but becomes broken, that is something does not go according to plan whether on the offensive or defensive side of the ball, assuming the quarterback still has possession of the ball, then s/he must improvise. No amount of contingency planning will prepare the team for all possible outcomes. In a desperate circumstance, the team might be prompted to try a move that they either have practised but have not yet attempted in a game situation or one which they have never tried before. Creativity is a part of every play, but different in this case. If they decide to go ahead with it then this becomes an instance of creativity in which something genuinely new occurs, hopefully for them in their favor. What they would then leave to the world is an example of novelty, which would have the effect of expanding their repertoire of plays that they can draw upon (should the coaches approve).

Novelty, though, does not take place with alarming frequency. Much of the world is called upon to repeat the past in an orderly fashion, and this is another area where God plays a role. In our analogy, there appear to be two sets of participants who have a status similar to divine overseers. The officials, for one, supervise the proceedings and place constraints upon the freedom of movement of the players. They ensure that the rules of the game are adhered to as the Whiteheadian divinity would enlist that part of the world which does not enjoy much self-determining power, ordinarily called the inorganic realm, largely to conform to natural law. This ensures that the world is dependable, so that one can be confident that it will behave similarly from one moment to the next. On the field, it could be said that the officials prepare the way for the following play by spotting the ball and signalling the next down to occur. This means that their placement of the ball gives the next play, or round of becoming, a specific spatio-temporal locus. Within all the constraints of the field of play and the rule book, the players are free to be as creative as they like (or as they dare).

The other set of divine role-players, as intimated previously, is the coaching staff. Aware of the talents, skills and abilities of their players, both their strengths and weaknesses, the coaches are best equipped to envision what strategies would have the greatest likelihood of meeting with success against a given opponent. Armed with this knowledge, they impress on the players an ideal for them to aim for and grasp. This is also how the process God deals with developing entities, by presenting them with an initial aim for them to consider in their own becoming. In their respective domains, both God and the coaches would be in the best position and have the clearest vantage point to make an educated guess as to what would constitute the ideal for these entities or players. Any gains that are made are then of benefit to all levels of the organization, athletes and staff alike. Both coaches and players, God and the world, reap the benefits from good play adoption and execution.

The coaching staff represents the realm of possibilities for the team. They alert the team about potentialities and make recommendations. The quarterback, however, if s/he is given the freedom to orchestrate the play in his or her way, makes these possibilities actual. Which play is ultimately selected is something about which the coaches will become aware only after it unfolds. Yet in a regular football game, the coach has much more authority than this and can pull any player in favor of another at any time if s/he so chooses. This is one way that the analogy breaks down. In the process scheme the coach would be powerless to make any changes. Rather, if the coach truly reflected the Whiteheadian divinity, s/he could merely persuade the quarterback to take one course of action over another but could never unilaterally coerce the decision. The range of influence on the part of God might be universal, but the extent of influence is limited. Hopefully the quarterback has an appetite for what the coach values.

Rounding out our discussion of the football analogy, spontaneity, interiority and mentality are most noticeably expressed in the following ways. As mentioned, a broken play can cause the quarterback to scramble and think on the run. The team under his or her direction awaits to see what s/he will do so as to salvage something from this play. Another instance is a fumble—a team can lose possession of the ball and perhaps even recover it on the same play. When a loose football bounces in unpredictable ways, the play book gets thrown out the window and all the players simply exercise their self-determining power to secure the pigskin (and hopefully training sessions have prepared them for such an eventuality). The choreography which ensues is frantic and highly entertaining, though perhaps not for the players involved.

We are now ready to examine the part of our map known as evolutionary theory and we will do so in light of our investigation of process thought. As a synopsis of the next chapter, let us say that evolution is a topic that has sparked much interest and debate ever since Charles Darwin's (1809-82) systematic formulation of natural selection in 1859. Some authors have described the subsequent relation of science and religion as conflictive and even combative. Yet the association of the two disciplines need not produce an unhealthy kind of tension. On the contrary, other researchers propose that nature and the evolutionary process it undergoes might also be understood as an arena for divine activity. This would have the effect of connecting science and religion, not dividing them.

CHAPTER FOUR

How It all Unfolded: The Later Years

PRIOR TO THE TIME OF ERASMUS DARWIN, Charles Darwin's grandfather, it was commonly understood that the natural world of biology, but not the physical world of physics and chemistry, was the arena of divine influence. The mechanistic picture of Sir Isaac Newton was believed to accurately reflect the physical world and to extend throughout the universe. But the natural world was thought to be reserved for injections of divine purpose.

Then at the time of Erasmus Darwin, evolutionary theories were already circulating. He along with others held to the concept of descent with modification. Erasmus anticipated future theories but he did not contribute substantially to them. It was his grandson Charles, whom he never knew, who systematized the findings obtained while acting as official naturalist aboard the ship named *The Beagle*. Charles formulated the idea of natural selection to account for the change in species over time. With the inspiration received from an article by Thomas Robert Malthus, Charles became convinced that competition for scarce natural resources leading to the struggle for existence propelled the evolutionary process into the survival (and reproduction) of the fittest—those with the best adaptive variations for the circumstances in which they find themselves. It should be mentioned parenthetically that the term "evolution" and the phrase "survival of the fittest" predate Darwin's publications and stem from Herbert Spencer.

We have Copernicus, Galileo and Newton to thank for removing humans psychologically from their presumptuous lofty perch atop the pinnacle or at the center of the universe; and then we have the Darwins to thank for demonstrating that all living things, humans included, are subject to the process of evolution and are thus not specially created. No longer are humans to be perceived as exempt from the modification which befalls all creatures. There is no fixity of species; they are not static but undergo change.

Yet is it proper to think in terms of the physical and chemical worlds as also having evolved? Let us take a closer look at this and recall a previous theme. In the initial picoseconds after the supposed big bang occurred, the fundamental forces of physics—namely gravitation, the strong and weak nuclear forces and electromagnetism, in that order—were established (Stoeger 1998, 166). That didn't take long! Once these forces and laws were set, what followed was the unfolding or development, which after all is what evolution means, of the cosmos up to the stage of the formation of galaxies and stars, when the only elements available were hydrogen (H) and helium (He). Large stars then became the factories where heavy elements such as iron (Fe) were forged from the raw materials of H and He (Stoeger 1998, 168-9). Hence the world of physics yielded the chemical, or the chemical world evolved from the physical.

This "inherent directionality" of the universe, it could be argued, was also evident at the next stage, specifically "the formation of planets." These and other stellar satellites became "the environments—the laboratories" where new and more complex molecules and compounds were manufactured, such as water, ammonia, and methane, from the five biogenic elements found in organic substances, namely carbon (C), nitrogen (N), oxygen (O), sulfur (S), and phosphorus (P). These elements are the building blocks for life, generating compounds like sugars and amino and nucleic acids (Stoeger 1998, 171). The two elements H and He comprise about 98% of all the matter in the universe, with another 1-2% devoted to chemically active elements that organize into groupings having biological affinities.

Then finally through a long evolutionary history we arrive at one of the products of this process, namely Charles Darwin, who reflected on it and schematized it. Regrettably, he was not able to offer a biochemical mechanism for natural selection since the work of the Austrian monk Gregor Mendel on the "factors" of inheritance was not known to him, thus the science of molecular genetics had to await a future generation. But this does not diminish the revolutionary impact of his work.

Darwin's conception of evolution by means of natural selection is often marked by the demise of the underfit or maladaptive individuals, which may lead to the extinction of entire species. Yet this process "has no foresight, nor does it operate according to some preconceived plan." Instead, it "is simply a consequence of the differential multiplication of living" forms. And "which organisms multiply more effectively depends on what variations they possess that are useful in the environment where the organisms live. But natural selection does not anticipate the future" (Ayala 1998, 107).

What is important for us to note about Darwin's proposal is that there is no aim, goal or purpose behind the evolutionary process if it accurately reflects a purely random undertaking. Evolution, according to Darwin and his disciples, is not headed anywhere. Despite their insistence on this point, fueled of course by the expectation that evolution remains a natural process without interference from outside, one can nevertheless detect a directionality in it. While it may not have an end in view, it seems undeterred in accomplishing at least one thing, namely an increase in complexity. One could argue that if left to itself the universe or perhaps even other universes like it would naturally be in the habit of fabricating the essentials of physics, chemistry and biology as our exercise indicates. Now whereas an increase in sophistication does not obtain across the board, for most species have either died off or remained relatively unaltered since they put in an appearance, an increase in complexity is undeniable where it does occur. Those few twigs on the evolutionary bush that do reveal this trend are unmistakable.

Physically, H and He describe the composition of most of the universe; chemically, the so-called inorganic world far outstrips the organic; and biologically, bacteria make up the greatest proportion of the biomass on the planet, even though they were the first form of life. (Is it more proper to claim that natural selection either commenced from or continued with microbes?) Other life forms then developed from this point and these amount to advancements in terms of complexity. What humans, for example, can achieve could legitimately be placed in this category. They are among the most adaptive creatures, they produce poetry and symphonies in addition to the laptop on which I compose. Does this not constitute an increase in sophistication? Scientifically, however, we are prevented from referring to this as "progress." Justifiably so, since the term is laden with liberal anticipations accrued from the nineteenth century, for this was the political climate in Darwin's setting. Ever since that time evolution has been associated with the notion of infinite progress, an idea that Darwin did not intend to convey. But to think of evolutionary developments, even if they are relegated only to those biological forms enjoying "advanced" status—albeit an admitted minority of species, as "complexification" would seem entirely appropriate.

There are still other factors that enter into the calculations, three of which seem to be most germane to our purposes here. First, there is debate as to the pace that the evolutionary process assumes. Some declare that evolution occurs through gradual changes alone, while others insist that periodic jumps called saltations, even if infrequent, mark certain modifications. The latter strategy has been termed "punctuated equilibria" by Niles Eldridge and Stephen Jay Gould, and a growing number of biologists maintain this "both/and," as opposed to an "either/or," approach in which "evolution leaps as well as creeps" (Birch 1998, 229).

Second, natural selection is the result of both struggle and cooperation. Peacocke puts it in these terms:

> The depiction of this process as "nature, red in tooth and claw" (a phrase from Tennyson that actually predates [Darwin] ...) is a

caricature, for, as many biologists have pointed out, ... natural selection is not even in a figurative sense the outcome of struggle, as such. [It] involves many factors that include ... more cooperative social organization—and better capacity for surviving such "struggles" as do occur (remembering that it is in the interest of any predator that their prey survive as a species!). (Peacocke 1998, 370 n.34)

Once again, an either/or method offers an incomplete picture.

And third, the organisms themselves are not passive components in the natural selection machinery. In Barbour's observation,

internal drives and novel actions of organisms can initiate evolutionary changes. The environment selects individuals, but individuals also select environments, and in a new niche a different set of genes may contribute to survival. Some pioneering fish ventured onto land and were the ancestors of amphibians and mammals; some mammals later returned to the water [perhaps after judging that sun-bathing was overrated] and were the ancestors of dolphins and whales.... In each case organisms themselves took new initiatives; genetic and then anatomic changes followed from their actions.... The changes were not initiated by genetic variations. (Barbour 1998, 421-2)

In what way could the genes instruct these organisms to alter their situation? In an interesting Lamarckian twist, he "was evidently right that the purposeful actions of organisms can eventually lead to physiological changes, though he was wrong in assuming that [these] changes occurring during an organism's lifetime can be inherited directly by its offspring" (Barbour 1998, 422). In any event, genes can only account for so much.

Darwin's use of the term "selection" might connote in the minds of some the presence of a purpose in nature. The metaphorical language he employed in his descriptions of natural selection implies a teleology and suggests the artifice of an agency, though not necessarily an outside source. In calling it a "power," he made it seem as though it functions actively and intentionally (Clifford 1998, 297-8). Darwin, though, did not wish to give the impression that these descriptors should be taken literally. Neither he nor his followers would understand natural selection as a conscious endeavor. Nevertheless, the terminology is suggestive. What can be stated, however, is that selection is in the direction of leaving more offspring that bear adaptive variations.

As a result, natural selection explains the design of organisms, at least those with adaptive variations. These survive and reproduce "at the expense of maladaptive" ones (Ayala 1998, 108). In addition to the regularities of natural law being operative here, chance also plays a major role. The mutations that generate these variations arise at random, "independently of whether they are beneficial or harmful to their carriers." Mutations can arise by virtue of exposure to radiation or to mutagenic chemicals in the environment, for instance. This random process is then "counteracted by natural selection, which preserves what is [presently] useful and eliminates the [presently] useless." Yet even here, that some variation might surface in an environment where it could flourish, and another not, is a risky proposition. In any case, without mutation "there would be no variations that could be differentially conveyed from one to another generation. But without natural selection, the mutation process would yield disorganization and extinction because most mutations are disadvantageous" (Ayala 1998, 109).

Evolution thus combines chance and law or necessity and places randomness alongside determinism (Ayala 1998, 109). Although seemingly counterintuitive, chance or contingency does not thwart the evolutionary process but actually augments it. Chance provides the means or the pool of resources upon which natural selection will act, thereby increasing the likelihood of the propagation of beneficial forms. Hence there is a commonality here with the quantum world. Recall that quantum mechanics suggests that there is a fundamental openness to the subatomic realm, such that indeterminacy becomes one of its properties. The particles contained in it are analogous to poorly defined clouds, in reference to their wave-like behavior, which makes them appear as though they are spread out over a region. If they did not have a particle-wave duality and were simply particulate in nature, their whereabouts could in principle be pinpointed with greater precision. But as it stands, we can only speak in terms of probability. Similarly, the evolutionary world is also marked by openness, where the future is "influenced but *under*-determined by the factors of nature acting in the present. In, through, and beyond the causal conditions we can describe scientifically, things just happen" (Russell 1998, 203).

If nature is authentically open, then there is no fully determined causal world. This would entail the inherent gappiness of nature, where there is no known reason, for example, as to why one nucleus of uranium decays in the next instant as opposed to any other. Admittedly,

> [a]mong large groups of atoms in everyday objects, indeterminacy at the atomic level averages out statistically to give predictable large-scale behavior. However, in some biological systems, ... changes in a small number of atoms can have *large-scale effects*. A mutation could arise from a quantum event in which a single molecular bond in a gene is formed or broken, and the effects would be amplified in the phenotype of the growing organism and might be perpetuated by natural selection. (Barbour 1998, 426)

(The outward appearance or form of an organism is its phenotype and its genetic complement is called the genotype or genome.) A microscopic change can therefore have macroscopic effects. Natural law alone would lead to stability in the form of repetition and rigidity, but "unpredictability would reflect indeterminacy in nature and not merely the limitation of human knowledge" (Barbour 1998, 426).

If we occupy an indeterminate universe, at least to a degree, then the world contains "natural gaps." These are openings in the causal regularities of the natural order. They reveal not holes in our knowledge but cosmic lacunae, which manifest themselves in the physical and biological worlds. Gaps would then be structural or systemic, not cognitive or epistemic. That a divinity could avail itself of such opportunities should not meet with immediate disapproval. God's active involvement in the world could then be understood as orchestrating, navigating or choreographing the unfolding of the already inherent potentialities of the universe into actuality (Peacocke 1998, 364).

This viewpoint, though, is not without its detractors. One objection to locating divine action in physics and biology is that this type of operation would still be judged as interventionist. In this format, God would step in to alter the course of events which would not otherwise occur in the way that God intends if left to itself. Were it not for the divine input, the desired result would not have been obtained. Another objection deals with quantum indeterminacy, for it can be argued that not even God could definitively know beforehand the precise outcome of a quantum event. If the result, at least in some small measure, must come as a type of surprise to the divinity, then this kind of hands-on approach may still not produce the effect which God intends (Peacocke 1998, 368 n.31). A third objection is that if "evolution is God's way of creating," then it betrays a strategy "that fail[s] to look very closely at the pain, the elimination of the weak, and the enormous struggle and waste involved in such a 'clumsy' process as evolution appears to be" (Haught 1998, 410 n.30).

These and other difficulties might persist while retaining a traditional conception of deity. Attempts to overcome these hurdles using classical categories of God, of the type that Thomas Aquinas would have endorsed, would ultimately lead only to a divinity that is a "determiner of indeterminacies," a "grand collapser of wave functions," or one that uses randomness or chance to bring about order. But this may not serve to settle the issues and the problematic aspects might remain. What may be called for is an alternate model for God and this is where Whitehead comes in. Process thought incorporates both science and religion. One could even say that it amounts to a bridge between the two. Whitehead's occasions of experience, like quanta, are discrete, momentary and indeterminate. He also drew from relativity the notion that events in spacetime are relative to their frames of reference, meaning that "all entities are constituted by their relationships" (Barbour 1998, 436). The world of entities is also evolutionary in that, both individually and collectively, they are marked by a continual process of development and becoming.

As for conceptions of the divinity, Whitehead's God functions persuasively and not coercively. Nor could it be otherwise even if God desired it, for the limitation on God's power is not voluntary but ontological and necessary (Barbour 1998, 440). This implies that the process deity's hands are tied when it comes to unilateral decision-making for these occasions of experience. But what makes process ideas most conducive to current science is that its God is not called upon to intervene so as to fill specific gaps in the world. This is because "God is already present in the unfolding of every event," though "no event is attributable to God alone" (Barbour 1998, 441). Otherwise entities would enjoy no power of self-determination. In fact, there are no gaps in Whitehead's metaphysic since God is involved in each round of becoming for all entities. God's contribution "cannot be separated out as if it were another external force, for it operates through the interiority of every entity, which is not accessible to science."

On the one hand, evolution in the Whiteheadian scheme proceeds according to conventional scientific wisdom in so far as entities enjoy self-determination. God exercises persuasive and not coercive power and hence can act only by luring, coaxing or enticing. Since God cannot force any issue upon the world and its creatures, not even God can know beforehand what entities will do with their freedom. Otherwise their self-determining power would be unauthentic and even illusory. Evolution will therefore take on a clumsy appearance since it is based largely on trial and error on the part of the entities involved. Yet evolution will also be different from the traditional approach in the way that God is involved in each event, for no entity goes without divine influence. This means evolution is not haphazard but can assume a God-oriented complexion. Formal and final causation can then be reinstated since God has an ideal in mind which God attempts to promote, and both God and entities have their own aims, ends, goals and purposes that they hope will come to fruition.

But the main point is that within a Whiteheadian framework God works not only with the evolutionary process but is intrinsic to it. God is part of the very same process; nature is part of God's make-up. The process God is naturalistic and thus also encounters evolution. Evolution and God, therefore, can be well within each other's comfort zones. What remains to be seen is the extent to which this world can be described as the product of design.

Does God Wear a Designer Hat?

Evolution does appear to display some type of directionality in that there seems to be a natural movement from the simplest of elements to structures that exhibit life. The unfolding of physics, such as galaxies; to chemistry, like atmospheres, oceans and the earth's crust; to biology, such as cell walls and membranes, is suggestive of a built-in drive of the universe to bring forth evolutionists. Our world represents at least one episode of the universe to "biologize," but there is at present insufficient warrant to declare that this universe is typically life-generating, for life may or may not be widespread in it.

The surface of the earth, separated on one side by water and/or atmosphere and on the other by the earth's crust, seems to be where the line is crossed, using the standard chemical designations of inorganic to organic levels of existence. This is where forms of life have arisen that boast not only cognition but also consciousness. The universe is now self-reflective. Some researchers interpret this set of circumstances as indicative of a plan in the works and a planner behind it. Such an architect is also allegedly an engineer, who has both an agenda and works toward its completion. Some might even be so inclined as to suppose that the culmination of the proceedings has been reached with the arrival of humans. For those who are convinced that this situation betokens a designer, they have referred to "it" as "God." The philosophical exercise geared to move from the intricacy of observed objects to a purposeful Director is known as the teleological argument for the existence of God. The additional anthropocentric notion that the universe is built so as to ultimately yield anthropologists is called the anthropic principle, which we have met up with before.

Given the presence of organic material in nebulae, comets, asteroids and other astronomical material, some of which deposit their contents like seeds onto a receptive earth, it becomes difficult to dismiss the idea that the universe is in the business of manufacturing biomacromolecules and ultimately life forms. One could even be provoked to detect a conspiracy here between physics and chemistry to produce the materials necessary and the conditions suitable for biology. But that biology should fashion humans is another matter entirely. The first could be regarded as self-directed, but can the second? Some may claim that once life has made its appearance it is inevitable that intelligent life should arise. Yet can we be so confident that this is the case?

The greatest height that so-called inorganic materials can aspire to is the self-replicating powers of crystals, like the silica comprising clay. This, however, does not produce variation but just more of the same. The next level of achievement is reproduction, and this is where natural selection enters into the calculations. As mentioned, the organic molecules necessary for life as we know it populate nebulae and the detritus of interstellar space. These molecules surface in the ordinary course of cosmic events and the steps leading to their appearance on the scene as well as their arrival on earth can be plotted with some accuracy. But the anthropic principle offers a different twist to the account. It is not enough for advocates of this conception to announce that the planet earth is a hotbed of biology. They also submit that the rise of life forms to include humans is already in the cosmic cards. The only remaining issue is which planet or planets get to host them.

Teleological arguments are not new—they have a long history, but proposals involving cosmic purpose have experienced a resurgence with advances in cosmology. The delicate balance of universal constants, such as gravity, electrical charge and so on, combine to make life possible, and this prompts certain researchers to maintain that someone or something must lie behind this sensitivity. If any of these factors were to be the least bit disturbed, then astrophysicists would not be around to comment on them. Those in favor of an anthropic reading of cosmic history submit that the universe, or if they are so inclined whatever lies behind or beyond it, had in mind all along that at a certain stage intelligent, self-conscious life in some form would appear on the scene. These ideas provide the motivation on the part of several thinkers to judge that either the universe itself, or who or whatever is responsible for it, had *Homo sapiens* on its agenda.

According to the late Harvard paleontologist Stephen Jay Gould's evaluation,

> The central fallacy of this newly touted but historically moth-eaten argument lies in the nature of history itself. Any complex historical outcome—intelligent life on earth for example—represents a summation of improbabilities and becomes thereby absurdly unlikely. But something has to happen, even if any particular "something" must stun us by its improbability. We could look at any outcome and say, "Ain't it amazing. If the laws of nature had been set up just a tad differently, we would not have this kind of universe at all." (Gould 1985, 395)

This is one of the problems with an after-the-fact argument—the fact itself is believed to indicate a reason for the fact in the first place. Yet there might exist no such warrant. Such facts may suggest but do not necessitate purpose. Purpose is one interpretation of life in all its intricacy, but it is not a logical requirement.

Nor does intricacy of present function entail purpose or design. Natural laws reveal no such intent. Citing Gould again,

> The fallacy of inferring historical origin from current utility is best expressed by noting that many, if not most, biological structures are co-opted from previous uses, not designed for current operations. Legs were fins; ear bones [in humans] were jaw bones [in reptiles] and jaw bones were gill arches [in fish]; incipient wings could not power flight, but may have served for thermoregulation. (Gould 1987b, 48)

Purpose in these cases highlights the fortuitous nature of current appropriateness in light of prior usefulness. If purpose is not apparent as usefulness in stages, then purpose is of no relevance between origin and final product. And if variations are not adaptive on route, then they will not survive until a protracted end point arrives. Thus if humans are the anticipated final product of a directed process, then purpose must have been evident all along the way for us to be able to speak of a purpose in the present. If this is true, then even the stepping-stones are valuable both at the time and in hindsight.

Proponents of the anthropic principle will need to decide at which point or points purpose is illuminated. If it is taken to occur at multiple spots, then purpose becomes periodic, but is it ever unbroken? The shorter the interval between purposeful points, the smaller the morphological increment to the latter phase, aside from intermittent saltations. This might render humans as just another purposeful stage on the way to something else. If so, then the anthropic principle will betray a tentative character and will need to be renamed. In any case, to think in terms of purpose as not truly evident until the onset of humans is to neglect the equally important exemplifications of purpose that would have led to humans. Purpose at the end entails purpose along the way.

Related to this teleological theme is the topic of progress. Evolution is said to embody a progressive movement toward higher biological forms. The steps are taken to proceed from the more simple to the more complex; as rungs in a ladder increase in elevation so too does biology assume ever higher structures. The trouble with this interpretation is that life is more accurately depicted as a bush with an enormous number of branches (Gould 1987b, 211) and "not a highway" leading to but "one summit" (Gould 1996, 21). There are a multitude of current outcomes in evolutionary lineages, together describing an explosion of forms and not a single orderly continuum. "Each lineage is a series of curious accidents with long periods of stability (or numerous variants on basic designs)" (Gould 1987b, 211). The idea that humans, describing only one lineage among millions, occupy the pinnacle position of biological forms casts the anthropic principle in a presumptuous anthropocentric light. Humans enjoy a set of variations, different from but continuous with previous ancestral forms, that have survived for a time and are currently adaptive. And while some environmental changes and successfully adaptive variations may be predictable, the onset of mutations giving rise to specific variations is not. Variations do not progress, they merely alter. There is no value judgment of progress attached to this random movement and Darwin admitted as much.

Humans are undeniably more sophisticated than the bacteria from which they allegedly arose. Yet "the earth remains chock-full of bacteria, and insects surely dominate multicellular organisms... If progress is so damned obvious," Gould retorts, "how shall this elusive notion be defined, when ants wreck our picnics and bacteria take our lives?" (Gould 1996, 145) Bacteria are supremely adaptive and have secured niches where they are firmly entrenched. Billions of years of competition have not extricated them from their long-standing position of ubiquity, nor have they shown a large amount of modification since their inception. Any judgment that progress has occurred must therefore "be construed as a broad, overall, average tendency (with many stable lineages 'failing to get the message' and retaining fairly simple [and steady] form through the ages)" (Gould 1996, 21).

Darwin did hold to "the phenomenon of increased complexity" but with the proviso that this be understood as an extreme case and not as a generally "pervasive feature in the history of most lineages" (Gould 1996, 197). Darwin actually disapproved of the use of the term "evolution" because of its progressive connotations. From the time that Herbert Spencer, the father of Social Darwinism, popularized it, however, Darwin reluctantly employed the term, though not until it appeared in his volume *The Descent of Man* in 1871 (Gould 1996, 137). His misgivings about the term center on a random process as possessing no built-in mechanism for advance. As mentioned, it would be more precise to hypothesize that views of progress can be traced more to nineteenth century British liberalism, which was the climate in which Darwin wrote and which of course has its own roots, than to an outworking of Darwinian evolutionary theory. If one is still insistent upon the retention of the concept of progress, then one will need to search for a framework other than the Darwinian one in order to buttress it against additional criticism.

If these reservations about the notion of progress prove to be warranted, then what are we left with? If there is no progress, then there can be no anthropos (human) as telos (final, or at least interim, product, aim, end or goal). If there is no telos, then no purpose; if no purpose, then no design; if no design, then no designer. Or is this series of rulings too hasty? Can God still be part of the picture?

If one of the roles God plays, or one of the hats God wears, is that of Designer, then there is every reason to suspect, in line with the conclusion drawn by Voltaire and using the phraseology of Leibniz, that this is not the best of all possible worlds. For one thing, it is unduly messy. The history of the world is littered with biological flops and failures—unsuccessful organisms strewn across the eons as, quite literally, evolutionary dead-ends. Evolution as ordinarily understood can be excused for its trial and error tactics of hitting upon, or more often missing, adaptive structures. But if a divinity were to be at the helm, one could question the course that has been charted. To rephrase a romanticist ideal, is it really better to have lived and lost than never to have lived at all?

If there is design and evolution is its method, then the objection could be raised that there is simply too much waste due to species extinction left in the wake of the evolutionary program. If this is the manner in which God intends the world to run at least up to this point, then one could say minimally that God tolerates waste. As eighteenth century geological figure James Hutton wrote, "the system of this earth has either been intentionally made imperfect or has not been the work of infinite power and wisdom" (Gould 1987a, 78). He was referring to geological forces, but his sentiments can also be applied to design. Gould envisions two options based on his reading of Hutton: "if things improve in time, then the world machine was not made perfect, and if they decline, then the earth is not perfect now" (Gould 1987a, 85).

Perfection, however, might be a philosophical category which neither God nor the world is required to live up to. The concept was imported from Greek thought into medieval theology. Nevertheless, the world does contain destructive elements, such as ice ages, which could drive one to conclude that such injurious natural impediments result in the world being detrimental to the longevity of organisms upon it. More will be discussed about this claim in the following chapter. At this juncture the question can still be raised as to whether this is what a God-designed world would or should look like.

A process alternative in the tradition of Whitehead offers an alternative that takes issue with classical representations of God's attributes. It begins by agreeing that God is good but rejects the viewpoint that God is all-powerful. Charles Hartshorne (1897-2000), another prominent figure in process thought, refers to the latter attribute as a theological mistake. A classical understanding is also distinct from a kenotic approach in which God places a self-limitation upon God's own power, thereby allowing the world at least partially to select its own path. In this case, God would have the ability to steer a different course but would not elect to do so. In the process strategy, in contradistinction to the other two, God does not have the inherent capacity to implement a plan without the input of the creatures. The obvious drawback of such a scheme is that there is no guarantee either that the ideal course of action will ever be adopted or even that God's method is the best way to get there.

For process thought, the physical and chemical worlds are largely though not entirely determined, while biology is not. Life forms exhibit self-determination, whereas physics and chemistry are extremely but not completely limited in this respect and are more relegated to repeating the past than are higher organisms. Physics and chemistry are, to a great degree, predictable; biology not as much. Biological forms can exercise more freedom than physical or chemical forms, but this freedom is relative and not radical or absolute. This makes Whitehead's God neither a designer nor a director of events as such in world history. Yet this God does have a purpose, though God cannot unilaterally actualize it. The process God has insufficient executive power so as to carry out the divine objectives in an unobstructed way.

In addition to purpose there is also progress in the process outlook, since God urges entities on toward God's ideal for them. This God desires that each entity experience a heightening of intensity in accordance with what it has the capability to strive for and integrate for itself. This hopefully will lead to the maximization of value and creative expressions that would usher in novelty. That these purposes of God are sometimes accepted means that the world encounters evolutionary progress.

In the evolutionary sense, God prompts the emergence of novel forms by way of variations in the hope that these structures will perpetuate the creative advance toward further novelty. The greater the ability to capture the significance of and respond to one's environment, the greater the self-determining power and, in turn, the greater the potential for novelty. The cycle of novel variations leading to greater response to environment and even more novel variations describes a processive movement that takes the shape of an advancing spiral. Yet there is no specific goal, whether ultimate or interim, in terms of envisioning a final form such as *Homo sapiens*. At the same time, natural selection does not occur randomly, for the process God has vested interests in certain outcomes over others. In spite of this, since nature is part of God, implying that God is partially natural, natural selection even if influenced by God is still natural by default. God is a Natural Selector but not Executor.

What the above means is that evolution is not Darwinian, for the process system involves the enticement potential of God's initial aims presented to occasions of experience. God proposes and the entity disposes, but God never has in mind an anthropos as a precise end product. Rather, God hopes that occasions will eventually achieve consciousness in some form. Now that this has come to pass, some entities find themselves in the position of assessing whether the process worldview, despite its shortcomings, is the most accurate solution to the mysteries which the cosmos presents to us.

In summation, 1) progress is not inherent in the Darwinian program but a concept imported from outside it; 2) if the traditional divinity is a purposeful designer, then the creation contains flaws which reflect poorly on this deity's architectural and engineering abilities; 3) the classical God may have had humans in view as an end product, but its methods are needlessly wasteful; and 4) given the present alternatives, only the process framework can adequately account for progress and purpose.

What's the Purpose?

There continues to be a sizable movement devoted to reviving the type of teleological argument referred to above. The approach known as natural theology, where the world that science investigates is used as evidence to substantiate claims that a divinity exists, incorporates cosmological and teleological forms of argumentation offered by such notables as Thomas Aquinas and William Paley. Exponents of this viewpoint hold that the intricacies of creation are indicative of purposeful design and point to a Creator.

As previously outlined, the forces of nature and its universal constants, like the velocity of light in a vacuum, the energy and rest masses of the proton, neutron and electron, the elementary unit of electric charge and the gravitational constant, are all delicately balanced. The smallest deviation from these physical parameters would have resulted in an utterly uninhabitable universe. For some reflective persons, the steps leading to a cosmos populated with creatures are so astoundingly fortuitous that it would be perverse to describe the cumulative effect of these events as merely the product of coincidence. Instead, the universe must contain some property of (allow me to coin a term) "anthropogeny"—the tendency for humans to emerge on the scene. The world, they would submit, is fashioned in such a way that organisms bearing self-consciousness would inevitably arise. To declare, though, that this faculty must come in the shape of *Homo sapiens* is the height of presumption and I therefore avoid it.

There are, however, non-theistic versions of the anthropic claim, Barrow and Tipler mentioning some of these. The difference between the two camps is that the theistic variety posits a Designer who had humans in mind all along and thus deliberately set up the cosmos to unfold accordingly, while the non-theists attribute the coincidences to the properties inherent in a fertile universe. The battle lines in the debate can then be drawn along the division between which natural processes should be interpreted as purposeful and which could safely be ascribed to ordinary mechanistic and hence purposeless sources.

For the imposing amount of factors that would need to be in place for humans eventually to reflect upon them, these extend into the biological as well as to the physical and chemical worlds. Specifically, the history of life on this planet is replete with the extinction of those species that could no longer adapt to changing conditions. It is the fate not only of all individuals to die but also of populations and species to become extinct. Adaptive advantages ultimately expire. Many mammalian species can expect to endure for five to ten million years before they must give way to another species bearing more adaptive variations. If the estimates are accurate, it has been roughly five million years since our earliest hominid predecessor, known as *Australopithecus*, flourished for a time (Gould 1980, 266). As for our species, *Homo sapiens* "arose at least 50,000 years ago, and we have not a shred of evidence for any genetic improvement since then" (Gould 1980, 83). "Anatomically modern" *Homo sapiens* evolved "a mere 150,000 years ago," but "[t]he real breakthrough in ... art and hunting technology occurred only about 50,000 years ago" (Morris 1998, 201). And if human civilization has been around for perhaps 40,000 years, then it has endured for about 2,000 generations, suggesting that it is still an unproven experiment. Nevertheless, if we correctly play the biological hand of cards that we have been dealt, then our species could have a long way to go, but there is no guarantee.

According to paleontologists, the earth has witnessed a number of major and minor extinctions during the course of organismic habitation upon it. Two of the main ones occurred approximately 225 (the Permian) and 65 (the Cretaceous) million years ago. The latter period—roughly a 500,000 year time frame, a mere geological instant—saw the extinction of the dinosaurs, ending their more than 100 million year campaign on earth (Gould 1983, 338-9, 346). Almost all the species to have ever existed on this planet have become extinct. That most other species died but our lineage was spared is highly improbable. If the demise of the dinosaurs can be attributed to a collision of an extraterrestrial object with earth, then we may have it to thank for allowing our lineage to flourish. This is a crucial point, for earth's biological experiments are not laboratory controlled, since there are outside variables to contend with—those hazards that come with living in a messy solar system. Only the best adaptations can hope to survive an extraterrestrial onslaught of that magnitude. Evolution does not prepare organisms for future contingencies. Those that did survive were the lucky ones, nor is this something that could readily have been predicted (Gould 1983, 329, 340).

Whereas a cosmos utterly devoid of life might have been the result of slight differences in the universe's initial conditions, once firmly established the world does seem to facilitate life. The various forms of life that do arise thrive for a while and afterwards go the way of the dodo—the first extinction (in 1681) attributed to humans since biological records were kept (Gould 1980, 281). Even with the appropriate universal constants in place, the odds are still stacked against humans as the bearers of consciousness. Humans and their ancestral forms were not halted in their tracks, rather their lineage continued. They were spared not because they boasted any inherent strengths or propensities in the face of, say, cosmic collisions. Extinction is the fate which all species must eventually undergo, yet some have been found to be more successful in postponing it than others. That some species have survived mass extinctions is due not only to some measure of adaptability, through the possession of a set of advantageous variations, it is also a matter of sheer good fortune, which is not the same thing as delicately fine tuned initial conditions and the coincidences of favorable constants. Indeed, this is a case of the downright lucky. The emergence of human forms is less likely than consciousness as an evolutionary outcome. Perhaps the world order would be understood better as "psychogenic"—that which brings about psychology, regardless of the package it comes in.

If I were a mystery writer who could observe the long stretches of evolutionary history and note the development of intricate features such as the ear from reptilian jawbones, I would be prompted to conclude that the organ could not have arisen by itself, that is by natural laws and forces alone, which do not know what they are doing. By this I mean that they do not conspire to orchestrate events so as to produce the new form. Natural selection has no such powers of foresight, but instead must have had an accomplice. I would imagine, as might a detective, that these occurrences are suggestive of a perpetrator who works in a premeditated manner. The process God could be a prime candidate for such a suspect. This divinity is no enforcer, but does present recommendations for the best courses of action to adopt. Such a deity would have the greatest likelihood of assessing what an ideal option might be.

The above suspicion would be reinforced when it comes to the topic of extinction. The trail of death and destruction that natural selection leaves behind it does not seem to be in character with a classical conception of God. Geological columns are littered with species that have become extinct. Now if the classical God is to be understood as having created a harmoniously functioning biosphere, then whence all of this extinction? Were these simply experiments that went awry on God's part? There appear to be two main possibilities: what occurs in nature either is or is not in conformity with God's will. In the second instance, God is displeased with the wastefulness of extinction. In the first case, God sees extinction as a tolerable by-product of an evolutionary process that culminates, at least to this point, in Darwinists. In the first, nature is accepted for what the future promises; and in the second, nature has God's disapproval for repeatedly having taken an ill-fated course.

If the latter possibility holds true, then God is the author of a creation that is defective and might perhaps seek to restore a lost harmony; and if the former, then God is more concerned with the finished product. Hence God is either tolerant of extinction and strives to correct it, sometimes even through responsible human management, or God has engineered a creation that works within acceptable but less than optimal boundaries. If extinction is looked upon negatively, then God's handiwork needs repair; if it is viewed positively, then God shows little concern for the weak, that is those species impoverished by disadvantageous variations. Both scenarios are unsatisfactory.

A way out of this impasse is to consider God as having neither the only or even the major hand in evolutionary history. The process divinity, over against the classical deity, does not mold inanimate, insentient material into responsive creatures. Instead, Whitehead's God lures an already relational cosmos toward making ideal responses. The fact that entities can respond in ways other than ideal leaves the door open for courting disaster. The calamity known as extinction may befall such creatures.

In the process view, this is the price to be paid for the potential of maximizing value, heightening intensity and advancing novelty. The process strategy is in danger of being wasteful but it is not needless, rather it is evaluated by God as an acceptable risk to take. The Whiteheadian method honors the creatures by safeguarding their (albeit limited) self-determinative powers. Since the entities of the process world relate and respond, this scheme is equipped to adequately account for the unfortunate onset of extinction. Nature is not defective in this outlook but is the fruit of creaturely responses to God's evolutionary beckoning.

As mentioned, Whiteheadian process thought does not describe a Darwinian situation, for nature does not select as such; rather, it is the arena for conducting business. The business that entities conduct is one of assessing the conformity of God's aim with their own and integrating what it can apply from God's ideal into their own becoming. God presents ideals for entities to strive for and grasp, they then seize upon their intended version of the future in as much as they are capable. The further task which God undertakes is likely the generation of novelty in the form of mutations in the quantum world of DNA base pairs, thereby yielding variations (some might object, though, that such a tactic presents us with an interventionist and hence anti-Whiteheadian coercive deity). Should they prove advantageous, the organisms bearing these adaptive variations can make good on their claim for a piece of the territorial pie, or niche as it is called. The Darwinian picture of natural history is neither directed nor progressive. This is contrary to Whitehead's view both of God and other entities, each a part of nature, directing their own efforts and partially self-determining their fates. Both God and entities have a purpose or design on their own outcomes. God's urging then constitutes a desired direction to evolution, and if appropriated, leads to progress. Whitehead thus supplies what Darwin cannot.

The next instalment of our journey will provide a look at several different scenarios of the role of evolutionary processes and how they have delivered us to where we find ourselves and the world today. This will also mark a shift in our method from mere presentation of viewpoints to a greater emphasis on interaction with them and ultimately a criticism of them. The figures whose proposals will be investigated are, in order, Stephen Jay Gould, James Lovelock, Rupert Sheldrake, Robert O. Becker and Pierre Teilhard de Chardin.

CHAPTER FIVE

Have We Made Any Progress?

USERS OF MAPS FALL INTO AT LEAST TWO CATEGORIES. The first understand traditional approaches to be the only ways to traverse the distance between two points. Other routes might be chosen, but they are all contained within the map as it stands. A new way then becomes simply a rearrangement or rerouting of the old. The second group consults the map as merely a guide to paths which have been taken in the past. But if it is judged that a better route might be adopted over previous ones, then the new way would be genuinely new. My suspicion is that the older paths have been exhausted. Newness might surface only through the use of a different map. Now this does not imply that the way will be without its hazards. I suppose that new ways, too, must be broken in.

The alternatives to be introduced here, subsequent to the first one, are by no means understood as firmly established in mainstream science. Instead, they are found, if anywhere, on the periphery of what is touted as legitimate science. Yet while there might be safety in numbers, I am persuaded that the majority opinion does not always reflect the most accurate position. To find oneself on the borders of a conventional map potentially indicates that the map's boundaries could benefit from expansion. We will consider whether such extension is warranted in these instances.

Stephen Jay Gould

If we are honest, evolutionary biology continues to be a speculative enterprise. Evidence of evolution taking place is not usually open to observation in the strictest scientific sense. This view is corroborated by biologist Lyall Watson (b. 1939):

> Not once in the history of biology has anyone been able to watch the simple differentiation of one species from another. So the most basic assumption of evolution, the existence of specific change itself, remains a theoretical construct. All that we have been able to do is to observe mutation within a species, and to assume that this is the mechanism which could account for the development of living organisms from specimens in the fossil record which look sufficiently like them to have been possible ancestors. (Watson 1980, 189)

Nevertheless, this does not detract from evolution as being methodologically fruitful and of boasting much explanatory power. But the fact that it is speculative means that it does not have a monopoly on all possible interpretations of the history of life forms on planet earth. We will determine if there is room for any competing approaches.

We are still on the topic of directionality and will remain so for the bulk of our study. Stephen Jay Gould is one author who opposes any notions of design and progress in the evolutionary scheme, yet at unguarded times lapses into mild acquiescence. While professing that "[l]ife arose at least 3.5 billion years ago, about as soon as the earth became cool enough for stability of the chief chemical components," Gould also states parenthetically that he does not hold to a position of the onset of life as stemming from "a chancy or unpredictable event." Instead, he envisions "that given the composition of early atmospheres and oceans, life's origin was a chemical necessity" (Gould 1989, 309). Regardless of how it unfolded, the hand that was dealt to the world included life and this was already contained in the chemical cards. Unlike Gould, though, whether one approves of it or not, life's emergence as unavoidable implies a direction to the proceedings.

Gould returns to his main biological thesis and emphasizes that organisms have not been designed for their environments; rather, the adaptive variations they enjoy, contrary to chemical necessity, came about through happenstance. The contingency of mutations leading eventually to beneficial forms is a random occurrence, and advantages in one setting or at one point might very well be disadvantageous in another. Moreover, those organisms best suited for a particular role might not be optimally suited; there may be room for improvement. This means they may be carrying traits that are redundant but have "a range of other potential uses by virtue of [their] structure—as we all discover when we use a dime for a screwdriver, a credit card to force open a door, or a coat hanger to break into our locked car" (Gould 1993, 118). The obvious difference, of course, is that coins, credit cards and coat hangers do not bear the forms they do because of genetic inheritance. They are products of human artifice which happen to, and this is Gould's intention, yield multiple uses.

In opposition to the doctrine of the fixity of species by an alleged creator, it would actually be detrimental if organisms were fine tuned to their surroundings, since environments change. If biological forms do not alter sufficiently along with changing circumstances, then extinction would ensue. Mutations provide the variations that natural selection, in a more deterministic manner, filters through its sieve. This process can supply the required adjustments for a lifestyle modification more conducive to the current setting (Gould 1993, 120). The important point to stress is that natural selection does not manufacture maximally adaptive forms for the environments in which they are placed. In opposition to the design proposal, Gould cites an example from the human body where "hernias and lower back pain are the price we pay for walking upright with bodies evolved for a quadrupedal life and not optimally redesigned" (Gould 1993, 369).

Another point Gould raises involves what he perceives to be an additional misconception about evolution. He recalls the old analogy of monkeys banging away on keyboards as the random way evolution is understood to operate in the popular mindset. It is held that in the long run, by the sheer number of combinations emerging, their energies will ultimately produce a work of Shakespeare. Gould maintains that this view is incorrect, for we will never arrive at such a final outcome if we must begin "each trial from scratch. But if we may keep the letters that, by chance, turn up in the right places and start each new trial with these correct letters in place, we will eventually get" a product which the bard himself would recognize (Gould 1987, 213 n.1). The trouble with this strategy, however, is the way in which we are to determine and be certain, in the midst of this messy undertaking, about which letters are the right ones—those to retain and those to discard—unless we have a standard by which to judge. Possession of such a standard would then of course preclude the need for our continued work on the project. And if this knowledge could be secured, then an end point to evolutionary development seems to be in view after all. If the monkeys can store and retrieve successful trials as required, then this "rightness" implies that a target is in fact aimed at or a goal worked towards and the language of directedness can justifiably be applied to it. To offer this analogy approvingly is to reinstate the notion of purpose and to rule on how closely the primates' work in progress approximates a final draft. Not only is it problematic to be privy to this knowledge—to come by way of it appears to be beyond human capabilities—but it would constitute, much to the chagrin of Gould, a progress report.

Gould later devotes an entire volume to his misgivings about progress as a function of evolution. He reiterates previous themes by stating that the mechanism of natural selection is inefficient in that it takes a long time to yield adaptations; indirect in that it acts on the raw material of evolution, namely variations, and does not supply them on its own but works with what it has been provided; and negative in that it eliminates and "remov[es] the illadapted, not by actively constructing an improved version" (Gould 1996, 221). In his analysis,

> Darwinian evolution ... is a story of continuous and irreversible proliferation. Once a species (defined by inability to reproduce with members of any other species) becomes separate from an ancestral line, it remains distinct forever. Species do not amalgamate or join with others ... they cannot physically join into a single reproductive unit. Natural evolution is a process of constant separation and distinction.

Gould does not deny that complexity in biological forms is on the rise in evolutionary history, but only in a "restricted sense," and "not as a pervasive feature in the history of most lineages" (Gould 1996, 197). He then turns to the sport of baseball to make his next biological point.

Among the great feats in this avocation, like Joe DiMaggio's 56-game hitting streak in 1941, Gould focuses on Ted Williams as the last player to hold a batting average of 0.400 or more in one season, and this occurred in the same year as DiMaggio's record. Since that time, the feat has never been equalled. Gould attributes this to the following phenomenon: "*As systems improve*" over extended periods of time, he contends, "*they equilibrate and variation decreases*" (Gould 1996, 112). Within the same system, "variation decreases steadily through time" (Gould 1996, 107), such that "[w]hen someone discovers a truly superior way," in this case 0.400 hitting, "everyone else copies and variation diminishes" (Gould 1996, 114). General improvement in play on the baseball field spells the demise of the exceptional. The bar gets raised so that, in Gould's analysis, the average shifts closer to what was once the exceptional (Gould 1996, 119).

His is a compelling argument, but I wonder if there are other ways of utilizing baseball statistics so as to reach a different conclusion. In the first place, a trend in human abilities is one consideration, but does his biological analogy work with reference to evolutionary time scales? Perhaps this or any other sport, for that matter, has not been around long enough so that one might draw evolutionary parallels from it. Evolution is not generally noticeable in human life times nor in the two-century time frame that baseball has been played. In the second, the diminution of extremes should have revealed a steady decline over the decades since the inception of baseball, yet this is not strictly the case. One should have seen the greatest ranges at the beginning when baseball statistics were first recorded. How then can one account for it not having occurred until the 1890s? In the third, does Gould's reasoning apply only to the highest levels of the sport, namely the major leagues of baseball, or also to the minor leagues known as the farm clubs or single, double and triple "A" teams? In fact, the set of professional ball players is itself an extreme and is not truly representative of the entire human population. Does nature actually operate as Gould suspects?

In any case, why should the mark last set by Williams (and others) stand as the most appropriate evolutionary analog? Why not home-run hitting, for example? Babe Ruth's home run record of sixty in 1927 stood until Roger Maris eclipsed it by one, hitting sixty-one in 1961 (a useful mnemonic), although Maris had eight more games to work with in a slightly longer regular season schedule. Thirty-seven years later, this standard was shattered by not one but two players, and in two straight seasons to boot. Mark McGuire hit seventy home runs in 1998 and sixty-five in 1999, while Sammy Sosa hit sixty-six in 1998 and sixty-three in 1999. Barry Bonds then surpassed both of these marks by setting a new standard at seventy-three in 2002. Indeed, if home-run hitting is on the rise, does this weaken Gould's thesis?

If Gould's proposal is correct, then why would home-run hitting not be similarly affected as batting average? Or is it? In an attempt to answer this question, I took the liberty of consulting a baseball encyclopedia for the home-run hitting leaders in each of the two major leagues—the National and the American. I elected to begin with the time that Babe Ruth rose to fame as the premier slugger in all of baseball, roughly at the end of the First World War. I plotted the figures on two separate graphs—one for each league—and looked for a pattern. The trend I noticed in the National League (NL) was that the bulk of the top home-run hitters in general were positioned in the upper twenties to low forties range in terms of number of home runs per season. The range then jumped by about ten after the Second World War, from the upper thirties to about fifty. This of course increased dramatically by 1998 when a new record was set. Over in the American League (AL) the pattern is somewhat more complex. It starts off high due to the exploits of Babe Ruth, but declined to a low in World War II. It reached another high during the Mickey Mantle and Roger Maris era from the mid-1950s to the mid-1960s. Another drop occurred in the first half of the 1970s and from then on there is a general trend towards an overall increase (being more pronounced, however, in the NL).

The point of these two graphs is in what they do not exhibit, namely a decrease as with batting average. The implication which may be drawn from this is that if home-run hitting were taken to be the analog for evolutionary theory and not batting average, then one might be led to conclude that evolution does in fact reveal progress or at least an overall upward trend. Contrary to Gould's thesis, improvement in play leads to further improvements, perhaps even in their steady increase. Or maybe this is just a case of statistics affording multiple interpretations.

There is another author who also takes issue with Gould's objections to directionality in evolution, though the focus of his disapprobation is a previous volume by Gould (1989). Simon Conway Morris undertakes to present a different account of progress using the same material as Gould, specifically the Cambrian period of approximately 530 million years ago and the type of invertebrate life that flourished then. Data for these competing versions was obtained from the remains of those creatures in the Burgess Shale. This is an area near the town of Field in British Columbia, Canada, where fossil specimens from that time have been remarkably preserved. What Morris finds noteworthy about them is the significant modification which took place during the Cambrian period, for it was the time when vertebrates began to emerge.

Currently, the lowest form of animal life is the sponge and the most primitive form of vertebrate life is the lamprey. Morris enlists this information into the service of his argument about directionality. As he puts it,

> It does not seem impossible that the evolution of animals could have stopped at the level of organization represented by the sponges,... By the time flatworms appeared, however, it seems that there could be no turning back. By then the basic genetic architecture was fully in place. (Morris 1998, 153)

This means that a general trajectory had been taken and another abandoned. Moreover, animals characteristically display bilateral symmetry, or an anterior-posterior axis, as well as neural tissue together with "conduction of nervous impulses." With this in mind, Morris is driven to propose that

> the appearance of the nerve cell must be regarded as one of the great steps in the history of life. This is because one path of evolution is then set towards the development of brains, presumably intelligence, and perhaps consciousness. (Morris 1998, 152)

The implication is that once certain stages in evolutionary history have been reached, nature betrays its commitment to follow along a particular path. The route cannot be forecast in detail, but the stage is likely set for one eventuality as opposed to another.

Reflecting upon these notions, Morris asks whether "an apparently identical character present in two animals [arises] because they share a common ancestor, or is it simply because the number of biological options available to provide a particular function is severely limited?" (Morris 1998, 172) He underscores that both can be adduced as factors:

> while the ... evolution of whales is ... no more likely than hundreds of other end points, the evolution of some sort of fast, ocean-going animal that sieves sea water for food is probably very likely and perhaps almost inevitable. Although there may be a billion potential pathways for evolution to follow from the Cambrian explosion, in fact the real range of possibilities and hence the expected end results appear to be much more restricted. (Morris 1998, 202)

Morris further outlines that the manner in which an animal propels itself through a watery medium "will fall into one of only a few basic categories" (Morris 1998, 205). Time and again, he adds, "we have evidence of biological form stumbling on the same solution to a problem" (Morris 1998, 204). This phenomenon is known as *convergence*, where organisms of different lineages "often come to resemble each other" (Morris 1998, 202).

The language employed here is not one of an aimless evolutionary process but, to use a traffic analogy, the path becomes a lane of least resistance which actually leads somewhere. If one takes the turn-off to a destination, then one is committed, at least initially, to heading in that direction. Thus contrary to Gould, even though natural selection does not necessarily generate the best solutions, the same solutions to a problem might be adopted by multiple species. And in spite of Gould's insistence that evolution lacks direction, it does appear that once certain steps are taken, trajectories can be projected with some precision since the options become limited. Gould and Morris represent the two main camps in this debate, meaning neither interpretation of evolutionary history is the obvious choice. Yet if Morris is correct, then directionality can be a function of a purely materialistic cosmos with no reference to a divinity. The inherent abilities of matter would then be reflected in the advancement of forms to a somewhat predictable end point, given the right conditions. No appeal to an outside Director need be made.

The place at which we have now arrived marks our point of departure from standard approaches to science and religion. The authors whose research will be examined in the remainder of this study all work with a different map. Each presents an analysis that does not operate within accepted scientific canons and as such has not been championed by the conventional scientific establishment. Their efforts also resonate with Berman's reenchantment themes.

James Lovelock

> [T]he "purity" of science is ever more closely guarded by a self-imposed inquisition called the peer review. (Lovelock 1988, xiv)

The first theory we will investigate is the Gaia hypothesis of James Lovelock (b. 1919), who objects to the narrow-mindedness of the scientific community in the manner in which it dampens curiosity and stifles inspiration (Lovelock 1988, xiv). He and others like him find that only certain scientific questions and answers are allowed to count as legitimate and this could have the effect of preventing science from receiving valuable insights where it might not otherwise seek them. He is keenly aware of this mindset toward his own work, which is not generally greeted or received warmly. The major features of his controversial idea are outlined below.

Lovelock was a researcher with NASA for a time and when studying the atmosphere of Mars he was struck with the discovery that its composition was in a steady state of near equilibrium. This disclosure prompted him to submit that any planet whose atmosphere functions in this way is indicative of a dead planet. When the NASA probe landed on Mars, Lovelock's prediction was confirmed—Mars in its current state is devoid of life. Lovelock then turned his attention to the earth and uncovered an atmosphere far from equilibrium. In response to this he proposed the idea that the presence of life on a planet and the activity issuing from this life produces a situation of atmospheric imbalance or disequilibrium. Not only that, but each—life and its gaseous products—is required for the continuation of the other. Life affects climate which in turn affects geology. And for life to be a going concern, it needs an atmosphere in disequilibrium, and vice-versa. In this case an imbalance results in the right balance for sustaining life (Lovelock 1988, 71). So far there is nothing exceptionally contentious about his claims, but that will change presently.

Lovelock reasons that during the history of life on earth the composition of the atmosphere modified according to both the needs as well as the waste products of the organisms living on it. The early environment displayed a predominance of carbon dioxide useful for the life-forms known as anaerobic microorganisms. Anaerobic refers to those chemical reactions that take place in the absence of free oxygen. Anaerobes also typically excrete methane as a waste product. Later when the remains of these anaerobes became embedded in limestone sea beds, for example, the bound oxygen from the carbon dioxide in these creatures was released and rose into the atmosphere leaving the carbon behind. The atmospheric oxygen then created a different imbalance to the disequilibrium already in place during the time of the anaerobes.

Later still, ozone from the fusion of three oxygen nuclei into one molecule formed a protective layer around the globe, enabling aerobic life forms to flourish. What was once a sanctuary where anaerobes could thrive then became a haven for aerobes. A carbon dioxide atmosphere made it a friendly place for anaerobes, but as the shift to a greater concentration of oxygen occurred, the aerobes found a home and the anaerobes had to retreat to the seas and eventually the guts of higher animals where there is relief from free oxygen. The result of natural anaerobic activity was the production of an environment hostile to them, perhaps the first instance of pollution. What is beneficial for some is detrimental to others.

This brief sketch of geochemistry reveals the substance of that which prompted Lovelock to propose the idea of "geophysiology" as a basis for what he calls "planetary medicine" (Lovelock 1988, xvii). Physiology refers to the regulative mechanisms set up in any living organism and Lovelock urged that the same may apply to the earth as a whole. He pronounced the earth's atmosphere to be a regulative system and named it "Gaia" after the Greek term for earth. Conventional science, though, is unwilling to accept the earth's atmosphere as physiological, since this term is usually reserved for biology not meteorology, despite the parallels with other life forms.

Organisms in the crudest sense are a bag of chemicals surrounded by membranes. Correspondingly, the earth's atmosphere is filled with chemicals in gaseous form and is surrounded by "membranes" such as ozone. But the notion of animation disturbs the scientific community. Yet Lovelock does not waver from his convictions: Gaia, he writes, has "develop[ed] the means of sensing the temperature ... [*inter alia*], ... so that production could be kept at the right level" (Lovelock 1979, 25). Such a regulative mechanism drives him to declare that "an active control system, however rudimentary, by the biosphere may have been the first indication that Gaia had emerged from the complex of parts."

The biosphere is the world of living things and the biota are those organisms within it. These two "taken together form part but not all of Gaia. Just as the shell is part of a snail, so the rocks, the air, and the oceans are part of Gaia" (Lovelock 1988, 19). Lovelock was impressed with the idea that the "atmosphere [is] a biological ensemble, rather than a mere catalogue of gases," and then was persuaded that these gases comprise "so curious and incompatible a mixture that it could not possibly have arisen or persisted by chance" (Lovelock 1979, 67). This led him to the proposal "that the biosphere actively maintains and controls the composition of the air around us, so as to provide an optimum environment for terrestrial life" (Lovelock 1979, 69). As such, Gaia "has continuity with the past back to the origins of life and extends into the future as long as life persists" (Lovelock 1988, 19). He even goes so far as to suggest that if Gaia were a living entity, then all living organisms would constitute "parts and partners of a vast being who in her entirety has the power to maintain our planet as a fit and comfortable habitat for life" (Lovelock, 1979, 1).

Such a view is inimical to mainstream science, yet Lovelock continues to understand the earth in Gaian terms. The surface of the earth, for instance, is the result of "the very presence of life itself. This is in contrast to the conventional wisdom which held that life adapted to the planetary conditions as it and they evolved their separate ways" (Lovelock 1979, 152). Not only is his ecological notion opposed to contemporary science but this opposition can be extrapolated to much of the modern Western mindset. As Lovelock explains "[i]n Gaia we are just another species, neither the owners nor the stewards of this planet. Our future depends much more upon a right relationship with Gaia than with the never-ending drama of human interest" (Lovelock 1988, 14).

Writing a history of life on earth from an ecological perspective would take into account modifications in both organisms and environment. Factors influencing the habitability of the world tend "to induce the evolution of those species that can achieve a new and more comfortable environment" (Lovelock 1988, 178). This implies that if the planet is adversely affected by our activity, "there is the probability of a change in regime to one that will be better for life but not necessarily better for us." Homeostasis—the movement of a system toward a state of what constitutes for it equilibrium and therefore health—is the norm until something upsets the (im)balance. A condition of (dis)equilibrium remains until a "force causes a jump to a new state" (Lovelock 1988, 13), with or without the presence of humans.

As already intimated, Lovelock's model would also have the effect of revolutionizing our understanding of pollution, for in his scheme it becomes an inevitable consequence of life at work:

> a living system can only function through the excretion of low-grade products and low-grade energy to the environment.... To grass, beetles, and even farmers, the cow's dung is not pollution but a valued gift.... The negative, unconstructive response of prohibition [toward pollution] by law seems as idiotic as legislating against the emission of dung from cows. (Lovelock 1979, 27-8)

For Lovelock the concept of pollution is anthropocentric and perhaps even irrelevant (Lovelock 1979, 110). Since he first went public with this view he has softened his position somewhat, yet still maintains that it is a question of degree. His reasoning is that certain naturally occurring substances may qualify as pollutants given sufficient concentration. Some materials, even beneficial ones, exhibit toxicity only in larger doses while others are poisonous in any amount. The issue appears to center around the question of "hazardous from what or whose perspective?"

The concern on the part of some individuals might be the welfare of those in what is called the First World, but in Gaia's perception it is "a region that is clearly expendable. It was buried beneath glaciers, or was icy tundra, only 10,000 years ago"

(Lovelock 1988, vii). And as for the nuclear threat, the earth in its history was covered with radioactive fallout from exploding stars and even now shows signs of radioactivity. Detrimental to humans, radiation "is to Gaia a minor affair." In opposition to anthropocentrism, "Gaia theory forces a planetary perspective" (Lovelock 1988, xvii). What matters is not the health of a single species but the planet as a whole, and this is why Lovelock names his approach "planetary medicine."

Hence pollution is relative to those organisms adversely affected by it. With the onset of oxygen in our atmosphere, its gas form "in the air of an anaerobic world must have been the worst atmospheric pollution this planet has ever known" (Lovelock 1979, 31), for it wiped out a large proportion of these creatures. As mentioned, the only places of safe haven for them at this time of about 20% atmospheric oxygen is below the surface of the sea where there is a lack of free oxygen and in the guts of most animals. In the case of the latter, anaerobes remain there in symbiotic relationship with their host, each benefitting from the presence and activity of the other. The host supplies nutrients as well as sanctuary from oxygen and these microbes aid in digestion—the breakdown of foodstuffs. Oxygen is thus a boon to some but a bane to others. In actuality, oxygen is injurious even for aerobes. Our skin, for instance, protects us not only from harmful microbes but flesh constitutes a covering that ensures the safety of our internal organs from the ravages of oxygen. Free radicals of oxygen (unbounded by other elements or compounds) inside our system can cause further damage, for which we are counseled by nutritionists to ingest anti-oxidants. Ironically, that which lets us live can also kill us.

Two atoms of oxygen make up the compound of oxygen gas and three form the gas known as ozone. The latter acts as a protective covering for the globe from ultraviolet light emitted by the sun. In its proper setting it is of benefit to the biosphere, but interestingly nature itself "has been in the business of destroying the ozone layer for a long, long time. Too much ozone may be as bad as too little. Like everything else in the atmosphere, there are desirable optima" (Lovelock 1979, 16). Lovelock's appraisal of this state of affairs bears repeating at length:

> Ozone, they said, prevents the penetration of hard ultraviolet radiation that otherwise would keep the land sterilized and uninhabitable by life. This was a decent scientific hypothesis and a very testable one at that. Indeed it was investigated by my colleague Lynn Margulis who challenged it by showing that photosynthetic algae could survive exposure to ultraviolet radiation equivalent in intensity to that of sunlight unfiltered by the atmosphere. But this did not stop the hypothesis from becoming one of the truly great scientific myths of the century; it is almost certainly untrue, and it survives only because of the apartheid that separates the sciences.... The members of each discipline tend to accept uncritically the conclusions of the other. (Lovelock 1988, 168)

With or without ozone, life would persist and Gaia would survive.

In claiming, for example, that the reason "[t]he Earth was saved from dying out" was because of "the abundance of its water, and by the presence of Gaia, who acts to conserve" it (Lovelock 1988, 192), Lovelock has earned for himself the reputation of not only being controversial but, what is worse in the eyes of the scientific community, teleological. Another example of this tendency is the following:

> I suspect that the origin of Gaia was separate from the origin of life. Gaia did not awaken until bacteria had already colonized most of the planet. Once awake, planetary life would assiduously and incessantly resist changes that might be adverse and act so as to keep the planet fit for life. (Lovelock 1988, 76)

Branded as one holding views that involve purpose means "a final condemnation" on his theory. Explanations using aims "in academe, are a sin against the holy spirit of scientific rationality" (Lovelock 1988, 33). He notes that "regulation implies the active process of homeostasis," yet at the same time he declares "[i]n no sense is this intended as a teleology, meant to imply that the biota use foresight or planning in the regulation of the Earth" (Lovelock 1988, 102). Gaia circumscribes the globe and is bounded by "the outer reaches of the atmosphere" (Lovelock 1988, 40-1). As life affects its environment, so Gaia influences life by sustaining it.

Evidence reveals "the Earth's crust, oceans, and air to be either directly the product of living things or else massively modified by their presence" (Lovelock 1988, 33).

In spite of his maneuvering, it is difficult to avoid the conclusion that there is a teleological flavor to Lovelock's presentation, especially when Gaia, which he regards as "the largest manifestation of life" on the planet, is described using language such as an "automatic, but not purposeful, goal-seeking system" (Lovelock 1988, 39). Goal-seeking usually entails purposive behavior. Nor does Lovelock portray Gaia in divine terms. He denies that Gaia is "a sentient being, a surrogate God" though he does assert that she "is alive" and we are "a part of her" (Lovelock 1988, 218).

We introduce his work with the intention of demonstrating a current scientific theory expressed along reenchantment lines. It is the notion of a living Gaia which concerns us here, and to this end Lovelock does not disappoint. Yet he persists in employing language ordinarily reserved for conscious organisms: he presses for an understanding of Gaia as "stern and tough" (Lovelock 1988, 212). She works "for those who obey the rules, but," like a taskmaster, is "ruthless in her destruction of those who transgress. Her unconscious goal is a planet fit for life. If humans stand in the way of this, we shall be eliminated with...little pity." For Lovelock, it "seems that there is no prescription for living with Gaia, only consequences" (Lovelock 1988, 225). Even if his intent is only to speak metaphorically, he lapses into literal moments.

Another difficulty with the Gaia hypothesis is that it is near-sighted. It does not take into account the broad scope of factors in the solar system which enter into the calculations of life on earth, regardless if the entire planet can be spoken of in organismic terms. I am not referring to the sun as, admittedly, the main player here. Granted that it will burn off most of its fuel in about five billion years, at which time it will puff off an outer layer that will engulf the earth, incinerating everything on it. No one would expect Gaia to withstand such an onslaught. The earth could not be faulted for having a mere nine or so billion year window of opportunity for sustaining life. The problem actually lies much closer to home.

In addition to the earth being placed a convenient distance from the sun in a habitable zone for life, our nearest celestial neighbor also has much to say about our well-being. The moon's gravitational effect on the earth is responsible not only for the tides but for keeping the earth stable as it rotates on its axis. Without the moon's influence the earth would not have a constant angle of rotation but could topple and assume a much different axis or wobble and have no well-defined axis. Some astronomers suspect that this very phenomenon has occurred to the planet Venus, which does not have a satellite. Perhaps this lack of a steadying influence, if not also a collision with some celestial object, caused it to turn upside down, giving it the only retrograde rotational motion in the solar system. (The planet Uranus has turned approximately ninety degrees on its axis, but not because it has no moons.) Any major deviation on this scale would be detrimental to life on earth if four reliable seasons could not be maintained.

This leaves little room for error if life on earth is to thrive. Life would continue to flourish as long as the moon remains a satellite. It currently recedes from the earth "at a rate of about 3 cm per year (roughly, as fast as your fingernails grow)" (Barrow 1995, 146). While this is hardly breakneck speed, over the course of billions of years this pace would amount to distances great enough to release the moon from its orbit. The earth is approximately ninety-three million miles from the sun, and the moon is about one-quarter of a million miles from earth. The distance from the earth where the gravitational pull of the sun takes over from that of earth is just one million miles, or roughly four times the current separation of the moon from the earth. Even though each billion years would amount to only twenty-thousand miles further separation, meaning the earth is not in danger of losing its neighbor before the earth itself becomes incinerated, the greater separation can have a significant effect on the earth's rotation during its lifetime.

The concept of Gaia would then need to be expanded to the earth-moon system and not simply to our planet's outer atmosphere. It would be damaging to this proposal to consider the earth as a living entity without the dead moon, much I suppose as it would be to think of humans without the protective covering of hair, nails and outer layers of skin cells, all of which are dead. Thus Lovelock's idea that Gaia will always recover, rebound and nurture some form of life on earth no matter what is thrown at it (internally) during its lifetime is inaccurate if we lose our closest celestial neighbor (externally). Hence Lovelock's theory must be modified at least to include the moon as a "dancing partner" (Barrow 1995, 149)—a viable option since the behavior of the earth-moon system can readily be regarded as though it were a "double planet" (Barrow 1995, 126).

Yet difficulties for Lovelock's notion persist even back on earth, for our planet is actually toxic to itself. In the case of volcanic eruptions, the earth is self-polluting. As Paul Davies explains, "the peak of a mighty volcano ... belches forth immense clouds of noxious gas" (Davies 2000, 25). A similar type of activity can also be observed on the ocean floor where geothermal vents spew out materials hazardous to many, but not all, life forms. Lovelock would concentrate on the fact that Gaia is ruthless and that life forms adapt even to these extreme conditions. Life may even have been incubated around these vents, an eventuality which could be understood as the earth giving birth to creatures out of its watery womb. As for myself, however, I tend to focus on the fact that Gaia effectively pollutes air, land and sea, thereby militating against the life it is alleged to foster. Of what benefit is it to promote life on the one hand only to threaten it on the other?

Lovelock might respond by saying that Gaia could be interpreted as a human body which is constantly in the business of turnover—cycles of cells form in our bodies with the fate of death after they have served their purpose. But this is the usual life cycle of bodies and the cells they contain. Nature on the other hand is merciless; the world kills its own. If humans were to behave likewise, they would be diagnosed as pathological—at potential risk to themselves and others. Does Lovelock uphold such a global organism?

On the topic of metaphysics, Lovelock rejects the extent to which science displays a "materialist world of undiluted fact" and religion exhibits "undiluted faith," for each is "equally unacceptable to [him] spiritually" (Lovelock 1988, 217). He does uncover, though, common links between art, science and religion, making them "mutually enlarging." His attempt is to demonstrate "that God and Gaia, theology and science, even physics and biology are not separate but a single way of thought" (Lovelock 1988, 212). Hence his inclusion in our analysis as one prospective candidate for a road map of God and the new metaphysics.

As a final comment, Lovelock acknowledges the anthropic principle of both theistic and non-theistic varieties but distances himself from them. He is content with the properties which the cosmos reveals that enables life and therefore Gaia to emerge, but "react[s] to the assertion that it was created with this purpose" (Lovelock 1988, 205). There may be grounds, however, for supposing that Gaia might be complementary to the anthropic cosmological principle. Whereas the latter emphasizes the physical constraints for life, the former stresses those biological conditions optimal for regulating and sustaining it. Lovelock suggests not only that Gaia is the very mechanism responsible for this sustenance, but that Gaia itself is alive. Together, Gaia and anthropic theory could form a complete package if one is in search of a process for life and its propagation. Despite not favoring design, Lovelock's proposal could provide the means whereby evolution could exhibit directionality, whether with or without a divine component.

Rupert Sheldrake

Still another metaphysical map is supplied for us by the work of British plant physiologist and embryologist Rupert Sheldrake (b. 1942). He studies the morphogenesis, or development of form, of organisms from their earliest stages. Sheldrake was prompted to embark upon his research when he came across two scientific phenomena which do not fit the conventional canons of his field. The cumulative weight of these anomalies drove him to submit an alternate mechanism operative in the world. Those aspects of biology which for him break with the norm are outlined below.

The first enigma Sheldrake mentions is regulation. The manner in which simple and complex structures can organize themselves in such a way that their internal systems work together harmoniously was for Sheldrake the beginning of his dissatisfaction with standard scientific explanations. That developing organisms can coordinate the assignment of resources to limbs and organs so that they will become those very limbs and organs and not others is far from clear. Regeneration is a second conundrum which for Sheldrake elicits further discontent with stock-in-trade scientific descriptions. The problem is this: when certain organisms, such as salamanders, experience the loss of a body part, what remains of the body can reproduce it, though it may not be as complete as the initial structure. What impresses Sheldrake is that this occurs not only in the embryonic but sometimes also in the adult stages of particular species.

At this point, conventional science might appeal to genetics for an answer to these difficulties, but for Sheldrake these efforts fail to be remedial since he believes that there is something beyond genetics that is going on here. The trouble resides in the observation that "the genetic programme must involve something more than the chemical structure of DNA, because identical copies of DNA are passed on to all cells; if all cells were programmed identically, they could not develop differently" (Sheldrake 1987, 25). Yet neither regulation nor regeneration is adequately explained with reference to genetics: "morphogenesis could be *affected* by genetic changes which changed the means of morphogenesis, but this would not prove that it could be *explained* simply in terms of genes and the chemicals to which they give rise" (Sheldrake 1987, 49). Differences in morphology "cannot be ascribed to DNA *per se*," but rather to "pattern-determining factors which act differently" in each developing member (Sheldrake 1987, 42).

Put another way, DNA contains not "instructions" but chemicals; yet there is something operative here in addition to chemicals. If so, this would be contrary to the type of naturalistic account mainstream biologists would defend, that allows certain cells to become, say, arm cells and others leg cells, all the while having identical genetic complements. Conventional science may then elect to counter with the following: "At this stage the concept of the genetic program fades out, and is replaced by vague statements about 'complex spatio-temporal patterns of physicochemical activity not yet fully understood,' or 'mechanisms as yet obscure'" (Sheldrake 1988b, 86). Mechanistic science has been caught smuggling non-mechanistic elements into its scheme. Nevertheless, Sheldrake has a case only so long as advancements in the field do not shed light on these mysteries. Were understanding of these activities and mechanisms to be forthcoming, then Sheldrake would need to revise his claims.

Yet in our current setting, the problems for biology do not end there, but are extended to behavioral patterns in certain species. Issues of instinct and learning come into play, such as when treating spiders which "are able to spin webs without learning from other spiders," and the migratory patterns of birds. The latter is especially curious in those instances when offspring are left to follow their parents who have had a significant head start, only to meet up with them again in the new territory (Sheldrake 1987, 26-7).

There are thus limitations to physical explanations. The mechanistic theory of living forms demands a physico-chemical answer for the above phenomena, though one does not appear to be on the horizon. Even the current terminology which science employs to describe genetic ideas involving "programs" and "goal-directedness" is more metaphysical than physical. Sheldrake states the philosophical difficulty in this fashion:

> the attempt to account for mental activity in terms of physical science involves a seemingly inevitable circularity, because science itself depends on mental activity...since physics presupposes the minds of observers, these minds and their properties cannot be explained in terms of physics. (Sheldrake 1987, 30)

Sheldrake discloses three rival explanatory strategies for the foregoing: "Vitalists ascribe these properties to *vital factors*, organicists to *morphogenetic fields*, and mechanists to *genetic programmes*" (Sheldrake 1987, 25). Whereas mechanists would appeal to chance events as the driving force behind evolution, vitalists and organicists would insist that it is "due to the activity of a creative principle unrecognized by mechanistic science." Mechanists trust that the answer will eventually stem from material sources, but it must be deferred until such time. Explanations, they believe, will be contained within the conventional confines of efficient causality—ordinary cause and effect. But vitalists and organicists have no such confidence, so only these two remain as surviving candidates for Sheldrake.

In the vitalistic case, there is a spark or life-oriented force on the part of organisms. Here the organic world is qualitatively different than the inorganic and can accomplish what the inorganic cannot, namely engage its environment. This makes the organic active and the inorganic passive. For vitalists, this spark is known as *entelechy*—a term borrowed from Aristotle. Sheldrake is critical of this position because the "physical world and the nonphysical entelechy could never be explained or understood in terms of each other" (Sheldrake 1987, 53). He locates a "dualism, inherent in all vitalist theories," which cannot be justified in "light of the discoveries of molecular biology of the 'self-assembly' of structures as complex as ribosomes and viruses, indicating a difference of degree, and not of kind, from [say] crystallization." One further competitor for scientific explanation of these phenomena is information theory, but Sheldrake takes it to be irrelevant to morphogenesis since

> it applies only to the transmission of information within closed systems, and it cannot allow for an increase in the content of information during this process. Developing organisms are not closed systems and their development is epigenetic, i.e. the complexity of form and organization increases. (Sheldrake 1987, 61)

This leaves organicism. Sheldrake believes that the organicist approach holds the key to the understanding of the emergence of form and pattern. In similarity with vitalists, organicists would claim that there is a creative principle involved in organic forms that is unrecognized by science. Both introduce a type of causation that is more than the sum of the parts of the developing system and can be interpreted as goal-directed with respect to regulation and regeneration. But in distinction from vitalism, organicists would oppose the very organic-inorganic dichotomy. In addition, Sheldrake's appeal to morphogenetic fields reveals that fields can also be external to organisms, whereas vitalistic sparks cannot.

Sheldrake's scheme has implications for the fields of both physics and biology. In the latter case he notes the three-dimensional configuration of complex protein molecules such as hemoglobin in which the folding process they undergo is "'directed' along certain pathways" and perhaps even "towards one particular conformation of minimum energy, rather than other possible conformations to the same minimum energy" (Sheldrake 1987, 74). Sheldrake reasons that it is "therefore conceivable that some factor other than energy 'selects' between these possibilities and thus determines the specific structure taken up by the system." He then outlines the nature of this agency.

As for the former instance, Sheldrake draws from physics the concept of fields. In the Newtonian world of forces, fields arise in material bodies and "spread out from there into space," whereas in contemporary physics "fields are primary: they underlie both material bodies and the space in between them" (Sheldrake 1987, 64). This provides Sheldrake with a suitable analogy. He borrows the idea of morphogenetic fields from the American biologist Paul Weiss, who conceptualized it in 1925, to describe the agencies which cannot be directly observed but which are detectable, like gravitational and electromagnetic fields, in terms of their effects (Sheldrake 1987, 76). These fields have as their sphere of influence the morphology of chemical and biological structures. They order the shape of physical systems and "[i]n one sense they are non-material; but in another sense they are aspects of matter because they can only be known through their effects on material systems." These effects are "in conjunction with energetic processes [but] are not in themselves energetic" (Sheldrake 1987, 25). This is one field which Sheldrake thinks physics has not as yet uncovered.

In the example of hemoglobin folding, Sheldrake contends that this field acts so as to assist the protein in "finding" its final form (Sheldrake 1987, 85). The final shape it assumes is not through some type of random search but is much more rapid than this. The same or similar pathways are then followed in subsequent foldings, prompting Sheldrake to postulate that it is as though these pathways form an ever-deepening groove allowing for ready access to the final form every time the pathway is traversed. He refers to these grooves as canals and the process as *canalization* in a non-material landscape. Such "canalized pathways" he calls *chreodes*, a term coined by experimenter C.H. Waddington, to describe a "canalization of development" (Sheldrake 1987, 79, 138).

Sheldrake names his proposal the *hypothesis of formative causation*, where the morphogenetic field causes a final form to be reached. He admits, of course, that not all morphogenesis is determined by it, for some patterns do arise randomly and others by "minimum energy configurations" (Sheldrake 1987, 89). His point is that morphogenetic fields account for many forms and then only in conjunction "with the energetic and chemical causes studied by biophysicists and biochemists." But there is a difference between his approach and those of teleological and design strategies since the end, aim or goal in view here is taken on simply because it has done so perhaps countless times in the past. The more an organism assumes a certain form and the more often it has done so in the past, the more likely it will continue to do so in subsequent instances. In this sense there is a "presence of the past." The past imposes itself upon or resonates onto the present by virtue of the depth of the canal that has been carved to date in these pathways.

The "causal connection" in this formative causation is achieved through what he calls *morphic resonance*, which enables formative causation to proceed. Yet morphic resonance is "non-energetic," meaning it does not propel formative causation. The distinction to be made here is that morphic resonance "does not involve a transmission of energy," though "[a]ll similar past systems *act upon* a subsequent similar system by morphic resonance" (Sheldrake 1987, 99, 120). Sheldrake illustrates this process with an analogy drawn from the art of photography. When a number of photographic films of the same subject are superimposed, the common features will be highlighted whereas the averaging effect will result in the other features becoming blurred. Formative causation operates by reinforcing the morphogenetic fields of the common features of past systems along their chreodes through the resonance of the final form. This makes the past present:

> as time goes on, the cumulative influence of countless previous systems will confer an ever-increasing stability on the field: the more probable the average type becomes, the more likely that it will be repeated in the future. (Sheldrake 1987, 103)

One system's influence on subsequent ones then depends on its longevity: "one that continues to exist for a year may have more effect than one that disintegrates after a second" (Sheldrake 1987, 104). Thus morphic resonance has a stabilizing influence on forms: "The more frequently this particular pathway of morphogenesis is followed, the more will this chreode be reinforced. In terms of [an] 'epigenetic landscape' model, the valley of the chreode will be deepened the more often development passes along it" (Sheldrake 1987, 107). These effects are cumulative, meaning the influence becomes more pronounced with each passing occasion. For those grooves which have endured the longest, such as the hydrogen atom—of the order of billions of years, "the morphogenetic field will be so well established as to be effectively changeless." For more recent ones, there might be a limited number of pathways that can be traversed, and which one or ones will eventually be taken is what Sheldrake hopes to predict and detect experimentally.

Hence in Sheldrake's view heredity is not only genetic but includes an inheritance of morphogenetic fields "of past organisms of the same species. This second type of inheritance takes place by morphic resonance and not through the genes. So heredity includes *both* genetic inheritance *and* morphic resonance from similar past forms" (Sheldrake 1987, 125). And since his scheme involves the presence or absence and use and disuse of forms, this implies that there is room here for a modified Lamarckianism of the inheritance of acquired characteristics in addition to Darwinian/Mendelian inheritance. A Lamarckian type of idea can resurface once the corresponding mechanism of morphic resonance is taken into account. So for Sheldrake these are not competing strategies but theoretically can work in tandem.

Living in Resonance

Sheldrake offers two illustrations of what he assumes to be formative causation in action. The first originates in the physical sciences and concerns the synthesis of a new crystal. According to his hypothesis, no previous morphogenetic field will exist for a crystal which has never been synthesized. After the prototype has been produced, however, its form will influence subsequent forms by morphic resonance. And the more often this form is manufactured, the stronger will the influence grow and the more readily will it form crystals (Sheldrake 1987, 107). The greater the number of crystallizations, the greater the contribution to the crystal's morphogenetic field by morphic resonance.

The above is not without actual occurrence; the issue lies in the interpretation of the results. Chemists sometimes experience initial difficulty in synthesizing a new compound, but the procedure becomes easier with repeated attempts (Sheldrake 1987, 108). Mechanists tend to explain this phenomenon in terms of

> fragments of previous crystals "infect[ing]" subsequent solutions. Where there is no obvious means by which these seeds could have moved from place to place, they are assumed to have travelled through the atmosphere [as far as the other side of the planet] as microscopic dust particles.

Sheldrake mentions the further claim that these seeds are allegedly "carried from laboratory to laboratory on the beards or clothing of migrant scientists. [But] the effect still happens in the absence of any such identifiable carrier" (Sheldrake 1988a, 84).

Sheldrake believes that his is the more satisfactory explanation, one implication of which is that forms other than those currently existing are potentially producible, but they await a time when "suitable seeds ... [will] ... put in an appearance" (Sheldrake 1987, 108).

The second example requires some introductory remarks. As intimated, formative causation affects not only morphogenesis but behavior as well, including motor activity. Patterns of bodily movement will also tend to canalize and stabilize as chreodes. They will first appear as reflexes, next as instincts and then with repetition as habits. Instincts for Sheldrake depend on morphic resonance "from countless previous individuals of the same *species*," and habits on morphic resonance "from past states of the same *individual*" (Sheldrake 1987, 174). Learned habits then "depend on motor fields which are not stored within the brain at all, but are given directly from its past states by morphic resonance." If Sheldrake is correct, then the ramifications of his hypothesis for education are such that it should be easier to learn something if others of the same species, and especially those in your own family tree, have learned it before. In Sheldrake's words,

> The originality of learning may be absolute: a new motor field may come into being not only for the first time in the history of an individual, but for the first time ever. On the other hand, an animal may learn something that other members of its species have already learned in the past. In this case, the emergence of the appropriate motor field may well be facilitated by morphic resonance from previous similar animals. If a motor field becomes increasingly well-established through repetition in many individuals, learning is likely to become progressively easier: there will be a strong innate disposition towards acquiring this particular pattern of behaviour. (Sheldrake 1987, 183-4)

Having covered the preliminaries, this brings us to the details of our second example, this time drawn from the natural or life sciences. Sheldrake cites the work of W. McDougall at Harvard in 1920, who exposed rats to a watery maze. One line of these rats was trained to negotiate the maze so as to find its exit more successfully. Both trained and untrained (or control) lines were subjected to the same experimental conditions to test for inheritance of learned responses. Three possible outcomes exist for this test. First, "[a]n increased rate of learning in successive generations of both trained and untrained lines would support" Sheldrake's view; second, an increase in the trained lines alone would corroborate an unmodified Lamarckian position; and third, an increase in neither would reinforce the orthodox Darwinian-Mendelian outlook (Sheldrake 1987, 189). The outcome was such that both trained and untrained lines were affected, that is the successive generations of both lines were able to learn more quickly.

Despite McDougall having tested for a genetic component to the inheritance of learned behavior, his experimental results were negative for his own research but positive for Sheldrake's. An attempt to repeat these experiments was later carried out by W.F. Agar in Melbourne. About this effort Sheldrake announces that "[i]n agreement with McDougall, they found that there was a marked tendency for rats of the trained line to learn more quickly in subsequent generations. *But exactly the same tendency was also found in the untrained line*" (Sheldrake 1987, 192). This would seemingly vindicate Sheldrake's proposal.

If formative causation is an accurate description of morphogenesis, behavior and motor activity, it would affect not only scientific orthodoxy but also its methodology. Sheldrake contends that the hallmark of repeatability must undergo renovation if it is to remain a distinctive of science. This is because every time an experiment is performed, like the one with the rats, "it would not be fully reproducible by its very nature; for in attempts to repeat it, the rats would be influenced by morphic resonance from the rats in the original experiment. To demonstrate the same effect again and again, it would be necessary to change either the task or the species used in each experiment" (Sheldrake 1987, 194). If any alterations occur in the morphogenetic fields of the test subjects, then this will most likely be injurious to experimental repeatability. As Sheldrake indicates,

> experiments depend on the presupposition that the laws of nature are the same everywhere and at all time. But if fields

evolve, then certain kinds of experiments may be intrinsically unrepeatable ... [which] raises the embarrassing possibility that such effects may already be influencing the results of conventional scientific experiments. (Sheldrake 1988a, 85)

Sheldrake expands on Lovelock's vision and reasons that if the entire universe were to be understood as a living organism, "whose regularities depend on habits," then "mechanistic phenomena become a limiting case" (Sheldrake 1988a, 86). Sheldrake borrows from Charles Sanders Peirce (1839-1914)—the father of American pragmatism who was later to influence Whitehead (at least indirectly)—the notion that natural laws are best treated as habits which are reinforced through repetition. Formative causation would enlarge the vision of science by proposing that the more often a pattern is repeated, the deeper will habit become entrenched and the more the phenomenon will appear to be describable by natural laws. The expression "force of habit" is taken seriously by Sheldrake, who believes that it has its provenance in physics and its outworking in biology and psychology.

Sheldrake also acknowledges his indebtedness to Whitehead but only in passing reference. His hypothesis is noticeably Whiteheadian and reveals parallels with process thinking, one example of which will suffice here. Sheldrake would agree with Whitehead's statement that what persists or endures is not matter but form, though Sheldrake would add "and the fields that give rise to them." Whitehead would counter with the claim that all which exists is relationships and the responses of subjects to them. One difference between their schemes is that for Sheldrake the strength of a field does not diminish with time or distance, but for Whitehead what we could call the "impact quotient" of the past on the present, in terms of causal efficacy, fades. If Whitehead's reservoir or extensive continuum of past objects is sufficiently similar to Sheldrake's fields generated by past forms—two respondents have even suggested that morphogenetic fields "can be interpreted as having a function similar to Whiteheadian initial aims" (Henry & Valenza 1994, 42)—then some adjustments will be needed to sort out the discrepancy.

Despite encountering vigorous opposition, Sheldrake insists that formative causation is testable and he searches for ways in which experiments on it might be conducted. In so far as it can be tested, it must be counted as a legitimate contender not only for scientific status but perhaps even for a new paradigm. As a science, it would be confronted with the same types of limit questions as any other scientific enterprise. For mechanists, this would include the issue of the origin of nature and its laws, and for organicists like Whitehead and Sheldrake, the origin of morphogenetic fields. Admittedly, formative causation explains repeated patterns of form, behavior and movement, yet "it cannot account for the origin of these patterns" (Sheldrake 1987, 199). Science is not equipped to address these concerns; conclusions could be drawn, if at all, on metaphysical grounds alone.

Sheldrake does not discount the operation of standard scientific phenomena in his hypothesis. As he confesses, "genetic mutations and abnormal environments may well have provided the occasion for the first appearance of new biological morphic units" (Sheldrake 1987, 153). But mainstream science can take us only so far. According to Sheldrake, we are left with three alternatives: as in the mechanistic approach, the "origin of new forms could be ascribed ... to blind and purposeless chance;" or "to a creative impetus immanent in nature" where the world is already in possession of this inherent capacity; or "to the creative activity of an agency pervading and transcending nature," that is, both in and above or beyond it. The second option above is indicative of a vitalist scheme and the third of the organicist. Sheldrake does not commit himself to either the second or third option but flatly rejects the first.

Of the three main metaphysical alternatives open to us, the first is materialistic, which might need to undergo renovation if there are more types of fields for physics to contend with than previously suspected. The second is a vitalistic approach involving immanent creativities such as Henri Bergson's notion of an "elan vital" or life force (Sheldrake 1987, 208). Sheldrake is sometimes branded a vitalist because of his contention that something out of the ordinary functions behind biological forms. He turns the tables, though, on the mechanists with his assertion that they are the ones who import vitalistic elements into their scheme. As mentioned, Sheldrake unmasks the "purposive mindlike organizing principles" of current biology in its description of "genetic programs" (Sheldrake 1988b, 86-7). In the contemporary scene, "genetic 'programs' or 'information' or 'instructions' or 'messages' [now play] the role formerly attributed to vital factors" (Sheldrake 1988b, 87).

The third metaphysical candidate is a transcendent reality of the type an organicist might nominate. Sheldrake believes that there is room in his scheme for such a proposition. In this strategy, the findings of science can be described, though not explained, mechanistically as limiting cases. By itself, however, the mechanistic model of the world is, in Sheldrake's estimation, truncated. Formative causation would enlarge the former's vision, and with Sheldrake's contribution we note the makings of a reanimated view of nature (Sheldrake 1994, 79).

By way of commentary, I wish to make the following remarks. Sheldrake's theory deals with morphogenetic fields, previously conceived by other researchers, which concerns the development of form. The structure of an organism is passed on to the next generation in a modified type of Lamarckian inheritance. According to this approach, the greater the historical presence of a certain form, the more it wears a groove or canal in the field that gives rise to it and thereby strengthens or reinforces it. A shape becomes imprinted onto the background field that generates and sustains forms. And while an imprint might not be energetic, it is an imposition and as such is causal. But shapes are not all that the field influences, since the effect reaches also to motor skills and behavior. Where conventional science and Sheldrake are at odds is in terms of the perspective taken on the explanatory power of genetics. On this score, Sheldrake lines up in opposition to the tradition. He fashions his hypothesis because he recognizes an impasse, and for him this is the limit of genetics as a tool for explaining biological function. To reiterate, in Sheldrake's view, genetics cannot account for either regeneration—the ability of some organisms to replace and restore lost anatomical parts; or regulation—the capacity of some cells to differentiate and specialize in the right place and at the proper time, despite all cells containing the same genetic material.

For Sheldrake, genetics can neither explain these features of certain organisms currently, nor will it ever be able to do so. For conventional science, on the other hand, Sheldrake's position is similar to the god-of-the-gaps strategy, where God as a solution to a gap in our knowledge becomes increasingly squeezed out of those spaces as science advances and broadens its epistemological scope on the world. Science would understand Sheldrake's impasse as a gap that will eventually be closed. Indeed science has made strides in this direction, as there are now believed to be regulator genes that switch on and off at critical moments so that a sequence of biological events may occur properly (Barbour 1998, 423).[10] The more the gap narrows, the less Sheldrake has a case. Yet is there something about Sheldrake's theory that can be retained even in the face of this scientific advancement? Perhaps there is, although some may claim that this too is indicative of an epistemological gap that will ultimately be bridged.

As specific examples, spiders are able to spin webs and certain birds can locate their parents far away after they have been left behind—all without formal training. Technically, these would fall into the category of behavior, though they seem to transcend this simple description. Typical behavior includes activity like mating rituals and social organization, if any. But spinning webs and locating parents could be interpreted as abilities beyond ordinary behavior. In another example, termites, which are blind, are capable of constructing an arch through teamwork. Each company takes on half the task and then meets up with the other company halfway. Coordinated efforts are required for accomplishing such feats. These capacities are out of the ordinary; perhaps they simply should be called unusual abilities. Some might suggest that birds contain an internal compass affected by the earth's magnetic pole or are guided by the stars, but that alone does not communicate to them the specific whereabouts of their parents. And the capability of spiders to spin webs does not by itself cause them to possess the engineering acumen to produce webs with geometric precision. It might be stretching it to assume that DNA can account for these abilities when, after all, its function is purely to code for protein. Or have we underestimated the power of protein?

With reference to Sheldrake's crystal example, Lyall Watson is of the opinion that a Lamarckian description of events is also germane to the chemical world:

> At this inorganic level, Lamarck was right, acquired characteristics can be passed on; but this is true largely because crystals don't reproduce, they replicate. They produce exact copies of themselves carrying on whatever information they happen to have picked up. They are stable, they replicate with accuracy; but if any change is wrought in them by an environmental pressure, they copy this imperfection as well, incorporating it faithfully into their memories. (Watson 1980, 57)

Watson goes on to declare that this "is precisely what clay learned to do, and what it taught to the first complex replicating molecules cradled in the folds of its bed" (Watson 1980, 58). This could have provided the impetus for the emergence of biological reproduction from chemical replication, and Lamarck's hypothesis could, in some measure, still be applicable to both. And in connection with a driving force behind his mechanism, Sheldrake leaves open the possibility of a divine "formative influence." While it might be difficult to believe that forms can "yield fields," recall that like morphogenetic fields, magnetic fields arise from substances—matter generates them but they cannot be sustained on their own.

On the Lamarckian theme, and responding to Gould's illustration, we reiterate that Sheldrake offers a mechanism for a Lamarckian type of inheritance transmitted by repetition through morphic resonance, thereby strengthening the morphogenetic field and the likelihood of further repetition of the form. And if his reasoning is accurate, would it also have an athletic application? Perhaps those who would come to Sheldrake's defense may cite the example of Roger Bannister's breaking of the four-minute mile in 1954 as a case in point. Ever since this mark has been eclipsed, the feat has been accomplished by many others on numerous occasions. Even if this were to be attributed to superior coaching, training and nutrition, the fact that it has become commonplace allegedly reinforces Sheldrake's claim. It might also mean that we could expect Babe Ruth's home run record to be broken more often with the passage of time. Yet why would this not hold for 0.400 hitting in addition, as per Gould's thesis? If Sheldrake is right, we should see more of it, not less.

As a final instalment of comments on the work of Sheldrake, mention should be made of a study peripheral to his if only to highlight the distinction between the two. Biologist Lee Alan Dugatkin illustrates the prominence of cultural evolution in the development of organismic behavior, in opposition to the assumption that purely biological factors have the lone role to play. In his account of female mate choices in a variety of species, Dugatkin discovered that females will imitate the selections made by other females and he argues that this "imitation factor," as a recent book of his is entitled, cannot be explained solely on the basis of genetics. For those researchers who insist that genes are responsible for this behavior, Dugatkin declares that "[h]ow a gene complex codes for mate choice is rarely known, and where such genes reside is never known, but that such genes exist is almost universally assumed" (Dugatkin 2000, 31). Even when gene-based models are crafted in competition with his findings, Dugatkin maintains that "biologists had to admit that cultural transmission of behavior could have produced the exact same sort of behavior" (Dugatkin 2000, 50).

Dugatkin receives inspiration from a naturalist no less than Charles Darwin himself, who in a letter to A.R. Wallace in 1867 stated that he was "a firm believer that without speculation there is no good and original observation" (Dugatkin 2000, 87). Dugatkin adds that "[w]ithout a hypothetical structure to work from," any empirical endeavor is but "an arbitrary and meaningless" exercise. He therefore takes the liberty to speculate in a vein different from the norm. In this he is like Sheldrake, though Dugatkin remains within the accepted parameters of mainstream biology. Not so with Sheldrake. The difference between their approaches is that observation on the part of organisms is required for imitation to occur, in fulfilment of Dugatkin's thesis, whereas in Sheldrake's portrayal fields are responsible for the duplication of behavior even in the absence of observation. Of the two, Sheldrake would be the one allegedly engaged in a heterodoxical or fringe research program—at the borders or margins of respectability and credibility.

Robert O. Becker

Robert O. Becker is a medical doctor whose research interests overlap with those of Sheldrake. He too is dissatisfied with traditional approaches toward regulation and regeneration as ordinarily understood in biology, and uses his disquiet as a springboard for lines of research that were branded by the reigning scientific orthodoxy as unorthodox at best and heretical at worst. Becker spearheaded a campaign against the proliferation of human generated electromagnetic radiation (EMR) and field production—everything from refrigerators and ranges to radio and radar—because of the harmful effects they can have on life forms. His efforts, however, have consistently met up with resistance from conservative elements with vested interests in the maintenance of the status quo. In the course of his investigations he uncovered both the boon and the bane of EMR. As an M.D., Becker focuses on factors that augment the health of organisms as well as those which diminish it. He disclosed that certain EMR is actually conducive to the well-being and even healing of specific maladies while other EMR is hazardous. What is interesting to note is that those situations which fall into the former category are consistent with the frequency produced by the earth, about 10 Hz (short for Hertz, meaning cycles per second (cps)).

Organisms found on earth, Becker states, either "resonate at [this] same frequency or show extreme sensitivity to it." It "remain[s] supremely important for most life-forms, [and is] the primary frequency of the EEG in all animals" (Becker & Selden 1985, 259). (An EEG is an electroencephalogram—a device which measures brain-wave patterns and activity.) Remarkably, the 10 Hz frequency of earth "can be used to restore normal circadian [that is, operating on a 24-hour cycle] rhythms to humans cut off from the normal fields of earth, moon, and sun." Artificial environments have been set up to study the effect of extremely low frequency (ELF) fields on cells. It was found that the frequency range of 30 to 100 Hz "cause[s] dramatic changes in the cell cycle time. This interferes with [the] normal growth of [an] embryo and may tend to foster abnormal, malignant growth" in addition (Becker & Selden 1985, 263). The range 50 to 60 Hz was revealed as particularly damaging to health, for it is "the most powerful" range, that "fall[s] right in the middle of the 'danger band' in which interference with growth controls can be expected."

The veritable explosion of instruments and devices that produce EM fields at hazardous levels is a direct result of their industrial and military manufacture and commercial use. Becker declares that earth's "normal field is contaminated with our own EM effluvia" (Becker & Selden 1985, 263) which he names "electropollution." He concludes that "[a]ll life pulsates in time to the earth, and our artificial fields cause abnormal reactions in all organisms" (Becker & Selden 1985, 328). As he explains,

> [Two] dangers overshadow all others. The first has been conclusively proven: *ELF electromagnetic fields vibrating at about 30 to 100 hertz, even if they're weaker than the earth's field, interfere with the cues that keep our biological cycles properly timed; chronic stress and impaired disease resistance result.* Second, the available evidence strongly suggests that regulation of cellular growth processes is impaired by electropollution, increasing cancer rates and producing serious reproductive problems....
>
> There may be other dangers, less sharply defined but no less real. All cities, by their very nature as electrical centers, are jungles of interpenetrating fields and radiation that completely drown out the earth's background throb. (Becker & Selden 1985, 327)

The reason we broach the topic of Becker and his investigations is the noticeable overlap with Sheldrake's concerns. Both speak of fields: Sheldrake's come into existence with new forms and then persist if repeated; Becker's are conventional though unexpected in certain contexts, and their strength diminishes with distance. Yet Becker's use of the conventional is unconventional, and for this he receives grief from the scientific community. Becker acknowledges morphogenetic fields as a possible explanation for biological phenomena such as regulation and regeneration, though he makes no specific reference to Sheldrake. And like Sheldrake he also leaves the way open for other effects which his fields might produce or support, such as extra-sensory perception (ESP) and similar psychic occurrences.

If there is room for a divinity in Sheldrake's scheme, so too can there be in Becker's. Moreover, for those interested in the fine-tuning of the world, the additional factor of earth's pulsations as attuned to organismic health is suggestive of a further detail for a design perspective on the cosmos. Alternately, it could be indicative of a Gaian strategy to ensure that the earth continues to be habitable. Whether this would point to the presence of a Designer with such foresight is, of course, another matter.

Pierre Teilhard de Chardin

So as not to give the impression that Whitehead's is the only process perspective, we consider the view of the Jesuit priest and paleontologist Pierre Teilhard de Chardin (1881-1955) and examine briefly the map which he offered. Like Whitehead, who reserves discussion of God until the fifth and final part of his magnum opus, Teilhard does not broach the same topic until the epilogue of his major work *The phenomenon of man*, completed in 1941. Further like Whitehead, Teilhard understands entities, although he does not use Whitehead's term, to enjoy both a within and a without. That all matter, and not only all organisms, bear a subjective as well as an objective side is what science, for Teilhard, has failed to recognize. Science concentrates on externality to the exclusion of interiority but Teilhard emphasizes both. For him, all forms have both outwardness and inwardness—an aspect of material reality which does not just appear somewhere on the hierarchical scale of creatures but not below it, or at some stage in evolutionary history and not prior to it. Although Teilhard does not disavow an organic versus inorganic distinction to the world (Teilhard 1965, 87), he does maintain that even what we might ordinarily be inclined to describe as 'dead' proteins comprising 'living' protoplasm in cells already possess "some sort of rudimentary psyche" (Teilhard 1965, 85).

Two of his biographers expand on these themes and outline the classification scheme Teilhard employs for the types of energy at work in the universe. He subdivides them into two categories: the first he calls "tangential" energy and in it he places the standard forces with which the sciences are familiar; and the second he terms "radial" energy—"a still unmeasured ... psychic energy" that drives elements upward from a state of lesser to greater order (Lukas & Lukas 1977, 169). The first describes external interactions and the second internal. In the words of these two biographers,

> From whatever "corpuscle" (or grain) is the tiniest, simplest inert physical unit, all the way up the scale of evolution to that marvel of complexity, [humanity], this increase in the "interiority" of things, Teilhard insisted, could always be seen as directly proportional to their complexity.

Radial energy is the cause that drives the complexification of organisms upward on the evolutionary hierarchy—Teilhard here endorsing a concept, as we have seen, of which Gould would disapprove. For Teilhard, science "confine[s] itself to examining only energy's 'tangential' aspect" (Lukas & Lukas 1977, 325).

The greater the complexity the more acutely will the organism's sense of awareness develop, which eventually leads to consciousness or soul (Lukas & Lukas 1977, 170). The appetite for even greater awareness ultimately gives way to the reflectivity of self-consciousness (Lukas & Lukas 1977, 171). This yields a different perspective on evolution:

> For [Henri] Bergson, evolution was a process by which things *diverged* into a multiplicity of forms. For Teilhard, it was a *conver-*

gent process—[one] of emergence from the darkness of unknowing into light. From the moment that reflection was achieved, the question of a goal appeared—a direction to be followed for those beings [capable of making the journey]. (Lukas & Lukas 1977, 172)

This sentiment places Teilhard squarely in the teleological stream or camp of evolutionary thought—where the majority of biologists do not reside. In his considered judgment, entities in the inorganic world boast an "inner reserve" or "functional capacity" as do those in the organic. This is similar to Whitehead's assessment, the difference between the two categories being one of degree and not kind as organisms complexify, thereby implying that the organic-inorganic distinction should be dropped. What is more, this means that the division between the physical and natural sciences should further give way to various subdisciplines all under the umbrella of biology: "matter, besides being polarized mechanically particle-to-particle, was also polarized from above" (Lukas & Lukas 1977, 325).

In Teilhard's view, the universe is a single organism whose properties are expressed in some measure at all scales, present in the electron as well as the elephant. Hence "psychic activity which manifests itself fully in human thought must also prolong itself backward to the point where the roots of matter 'disappear from view'" (Lukas & Lukas 1977, 168). All of existence is corpuscular or granular in nature and enjoys inwardness, although at the lowest levels there are "only 'fleeting hints' of it." Thus "given the required conditions," no matter is disqualified from "evolving toward spirit." In addition to being eligible to climb higher on the scale, humans could also move closer together. With cultural evolution in mind alongside the rise of technology and global communications, not only does the earth have a covering of life on its surface, it also boasts a "thinking skin." The image Teilhard draws for this recent acquisition is humanity as potentially "becoming one single organism with a single nervous system, tightening its hold upon the planet" (Lukas & Lukas 1977, 90).

For Teilhard there is a definite progression of events as the cosmos unfolds. The first stage is a geogenesis or world-formation from physical structures and chemical elements. The second is a biogenesis, where life arises out of pre-life and results in a biosphere of living forms. The third is a psychogenesis, where the advancements or promotion of living forms results in those organisms with nervous systems, culminating in its human expression. The fourth is a noogenesis in which mind or consciousness develops from nervous systems. In this last phase, thought expands to the point where the earth can be said to boast a mental membrane or thinking layer, which Teilhard calls its noosphere (Teilhard 1965, 164, 201). His notion of the within of things implies that the consciousness enjoyed in the fourth stage was already present in the first, or as he puts it, "emprisoned in the matter of earth" (Teilhard 1965, 78). But the progression does not end there. Rather, in the course of time all such mentality will be drawn toward a universal center or principle which he names the Omega point (Teilhard 1965, 288).

What is clear from Teilhard's presentation is that evolution most assuredly has a direction and does not involve chance alone (Teilhard 1965, 121, 161). He even refers to the process as having a "specific target," "a precise *orientation* and a privileged *axis*" (Teilhard 1965, 121, 157). In effect, what obtains is a type of cooperative relationship of Darwinian/Mendelian chance and the "play of external forces," together with a Lamarckian psychical selection of these "strokes of chance which are recognized and grasped" by developing forms (Teilhard 1965, 165 n.1). For those who are not cognizant of evolution as headed anywhere, however, and this represents most biologists, the Teilhardian view would not have much impact.

One might be prompted to ask, though, if some aspects of reality in this progressive scale appear where they were not before, then does this imply that there was ever a time when physicality was without mentality? According to one respondent of Teilhard, neither matter nor mind can exist without the other—one cannot have either of them in a pure state (Haught 2000, 178). And unlike Whitehead, who focuses on a different *realm* from the one we inhabit, namely the mind of God from which possibilities are communicated, Teilhard stresses another *era*—the future, toward which the Omega attracts and beckons the universe forward (Haught 2000, 84). But we need to decide which it will be: if there is a point toward which the universe is drawn—centered on Omega with God as the center of centers—then evolution has a direction and God is at the helm; if the future is indeterminate, as with Whitehead, then it does not do any good for God to have a single point in God's cross-hairs, for the self-determining power of creatures might cause them to deviate from God's intended ideal and miss the target. One cannot have genuine openness alongside an anticipated destination. One can only choose between a vision and a plan—the former is hoped for, while the latter is deliberately enacted. Moreover, one might ask whether the evolutionary process can be evaluated adequately and properly by those within it, or need there be a more objective frame of reference from which to judge (if there be such an available vantage point)?

Indeed, if we are in the grip of and are drawn toward an absolute—the Omega point—then how open can the future actually be since it cannot deviate from this point? The paths that are traced out might differ slightly, but there is limited freedom of movement on them and little margin for error. And with respect to the capability of matter to rise spiritually, I suspect that not much hope can be held out for the bulk of matter in the universe. Teilhard admits that the right conditions must prevail, but where might these exist other than on earth? Even on our planet, in the majority of cases there would be little chance to advance to this extent.

Becker also makes reference to Teilhard regarding the noosphere as an "ocean of mind, arising from the biosphere as a spume" (Becker & Selden 1985, 270). The human body's own electromagnetic capacities could then provide access "to the great reservoir of image and dream variously called the collective unconscious, intuition, [or] the pool of archetypes" of the kind studied by Carl Gustav Jung. But the fields produced by industry and the military complex would effectively drown out this noosphere, for if it exists, then "our artificial fields must mask it many times over, literally disconnecting us from life's collective wisdom" (Becker & Selden 1985, 327).

In an attempt to bring this chapter to a fitting close, if we are to accept Teilhard's approach, then Lovelock's Gaia would survive but could not think, by Becker's reasoning; Whitehead's world would survive but could be in danger of missing the Omega point, by Whitehead's own reasoning; Sheldrake's fields are too past-oriented to work toward a future Teilhardian culmination; and Becker's fields are in tune with those of earth and therefore Lovelock's Gaia, but the artificial fields of humans would bring the Teilhardian evolutionary progression to a halt at the level of noogenesis by preventing the noosphere from remaining intact. My suspicion is that each of the five standpoints would be placed in a somewhat negative light when seen from the perspective of any of the others. So which of them should be the standard against which all the others are appraised? Or is there another model that we have not as yet considered? Is our map still helpful and should we forge ahead or simply be let off at the next stop?

CHAPTER SIX

New Vistas

I FIND MYSELF HAVING RESERVATIONS about the Whiteheadian framework, and since we are sometimes using it as a basis for comparing other viewpoints, it is worthwhile checking to determine if we are applying the proper grid. I call them reservations but could just as easily refer to them as objections and have done so elsewhere.[11] Yet for the time being I will use the milder term since the example I will provide does not evoke a knock-down argument, as though convincing everyone who comes across it of the weakness within process thought in this regard. Whitehead was too careful to let that happen. But was he sufficiently thorough so as to completely circumvent any fragility in his system? Here is the source of my disquiet.

The Past as Causal

The past is usually understood as being inactive and can no longer have an impact as such on the present. And if human nature is any indication, the repetition of historical errors reveals that the past supplies an insufficient reminder that unwanted consequences from previous actions might reappear if one does not heed the warnings. We may know what to avoid, and in this sense learning from history does occur, but contemporary circumstances often seem to overpower the benefits gleaned from prior education. Somehow we might think of the current situation as different in type so as not to fall within the same bounds as a previous failing or mishap. Once encountered, however, we become painfully aware of its congruence to the past. But by then it is too late. We discover that our learning has had a "best before" date.

On this score, the past does not impose its wisdom in any automatic sense. The present is the only scene in which efficient causation appears. If this is true, it would be detrimental to the process approach. For Whitehead, the past truly exerts efficient causality. To be settled does not mean to be without effect. It is unlikely, after all, that the cumulative weight of the inherited past could effectively be avoided. To non-Whiteheadian ears, though, this notion might sound counterintuitive. How is it best to read the past, then, when we recognize that we are products of our upbringing, we have a socio-political-economic legacy to contend with, we may be haunted by past decisions and we bear the benefit and/or burden of genetic inheritance?

If we are alert enough to notice what we have just asked, it would appear that our past very much influences how we presently conduct ourselves. If we are affected in all these ways then there must be something which affects. Some persons might submit that it is only our psychology which produces those kinds of responses and that they are fetters from which we ought to become unshackled. Our makeup is geared toward present responses to what has gone on before in our lives, and the past can and does impinge upon the present, but it can also be deflected. Yet this is precisely Whitehead's point. We receive from the past and respond to it based on the degree of self-determinative power we enjoy.

Our memories, for instance, are the present distillation of our past occurrences. We know our memories most intimately; no one else is privy to them, unless we communicate them, and even then they are not known as we know them. We experience our past through its present representation in our recollection of it (which might suffer in its accuracy with increasing age). Through our memories, though not exclusively, we are related to our past, albeit imperfectly. The filtration system of our interpretive schemes sees to it, for example, that the evidence of witnesses in a court of law is not always dependable. Our memories are not an accurate reflection of what has happened to us, but the way we currently view what we previously experienced. We do not reexperience the initial experience but a revision of it. In this way it undergoes a modification over time. In this way also, the present acts upon the past and edits it. We shape our past and are also shaped by it. This change of emphasis is different from Whitehead's intention, although he might have taken it into account.

What remains implausible to the thinking of some people, is the idea that the past can be forceful. But one need only be reminded of an event that one would just as soon forget, to realize that the past can still pack a punch. Perhaps it is true that the past does not have the power to initiate something so that a process might get underway. This is where Whitehead's God would come in. God initiates an event by giving it a spatiotemporal locus, or an address, and an ideal around which the occasion's efforts might be centered. This denotes the initial input of information into the system. Beyond this, the occasion draws data from the past as to what it was, integrates them into an understanding of what it is and generates the shape of what it can become. The movement from what it was, is and becomes marks the passage from subjectivity to objectivity.

This process also involves different modes of perception. In an age when information received through the senses is believed to be the foundation for what can be understood about the world, Whitehead heralded another perspective. In opposition to the trend championed by the British empiricist school, including such notables as John Locke and David Hume, Whitehead maintains that sensory perception has a more fundamental level from which knowledge may be obtained. He names this perception in the mode of causal efficacy, which appropriates information from the past in an unprocessed or prereflective and pretheoretical way. It is the initial impression brought about by an object in one's immediate past environment. Whitehead's argument, contrary to the view of Hume in particular, is that objects are emotively felt before they are sensed. The next level is termed perception in the mode of presentational immediacy. These two modes of perception are pure, for Whitehead, but the latter is more sophisticated and complex. According to one process thinker,

> Whereas causal efficacy, the mode of inheritance from the past, transmits into the present, data that are massive in emotional power but vague and inarticulate, presentational immediacy transmits data that are sharp, precise, spatially located, but isolated, cut off, self-contained temporally;... (Sherburne 1981, 236)

Presentational immediacy presupposes causal efficacy and elaborates on it. It adds definition to the former perception and makes it more relevant. The first notice one takes of a lake, for instance, might be "big wet," and only subsequently might it register as a body of water on which one needs to navigate. To this end, perception in a third mode, the mixed or impure mode of symbolic reference, is what is usually known as sensory awareness. Whereas in presentational immediacy something registers out of the corner of one's eye, so to speak, symbolic reference adds to it the qualities and properties characteristic of the object in question. Science operates at this third level. This is where both the prior modes are synthesized and integrated, and where empiricist views commence.

An analogy for the above can be drawn from human physiology. When one experiences a pain, say a burn on a finger, it is first registered in the peripheral nervous system (PNS) as a dull ache. But shortly thereafter the pain becomes registered through the central nervous system (CNS). At this point the pain is sharper and more refined and a theory has begun to form in the beholder's thinking about its nature and onset. The culprit might be identified as a kettle or an element on a stove, and the finger is responded to empathetically in an attempt to nurse it back to health. The experience of this pain is analogous to the process understanding of an event. The progression in such an event is from emotional or felt exposure to conceptual encounter to considered response. Once complete, the process begins again, hopefully without further discomfort.

So far the above appears to be a confirmation of the process outlook, but is it enough to say that the past imparts to us such a force which cannot be ignored? Here is where my misgivings surface. Our inability to avoid the force of our past on us at present is like saying that we follow a certain trajectory from which we cannot depart. We, however, are not completely like a ballistic particle which, after being launched follows a relatively predictable parabolic path and displays inertia while doing so—it is unable to alter its state of motion. To say that we are similarly affected by our past may be little more than saying that we are creatures of habit—we have a certain routine and thereby exhibit specific behavior patterns. But if the past acts on us at each point, this is unlike anything ballistic. The object traces out a path but not because its immediate past forcefully places it there.

There is an event in this object's past that causes it to describe a pattern—that which launched it in the first place. Yet from that instant onwards, the position of the object at any subsequent point does not come about by its immediate past as such. All pasts except the initial launch are inertial—no position imposes a force upon the next position save the launch point. All the intermediate points between launch and landing are ineffectual in imposing a force on the next point in the path since inertia prevents them from contributing anything more than the effects of the initial launch. For the particle, the recent past has been effective, but the immediate past is a lame duck. No recent past exerts a force any longer; we see only the effects of a more distant past. The launch is responsible for the location of the particle at any point, disregarding for the moment other factors such as friction due to wind resistance and so forth, but it exerted a force only once and not at each point.

Humans, though, are not like these particles since we are not inertial—we can modify our trajectories and alter our state of motion. We encounter a series of "launches" in our lives, but like the particle, they do not act on us at every point, they only help to explain where we are at a certain time. We are shaped by our past but it does not rule our present, only present forces could do that. The past does not impose present forces, only the present can perform this feat. Both we and the forces are in the present and come from the present. We may take our cues from what has gone on before and be informed by it, yet this is not the same as an active past. To the extent that there is continuity of identity, we are connected to our past, but present interpretations of past events, not the past as such, influence present responses.

In the ballistic example, the cause of a particle's position can be linked to the impetus it was originally given and the trajectory it is subsequently tracing out. The reason it can be found at a specific point along its trajectory involves the position it occupied immediately prior to the one in question. The cause gives us a trajectory and the trajectory gives us a reason. The position of the particle immediately prior to the current one, now in the immediate past, is not the cause of the present state, rather it is more of an explanation. One can point to a past event or series of events that account for the present shape of things and in this respect we are passive; but our present encounter with it elicits a response and this marks a deliberate engagement with it—in this respect we are active. A proposition made to someone, for instance, is definitely efficiently causal since it demands a response, but of course the response is not contained within the cause as though it were waiting to unfold. Yet can that which is no longer active still be causal? The past is enlisted by the present; the emphasis is on the past being a recipient of the present and not vice-versa. The past potentially instructs, and it is also something to which the present succumbs.

I will, for instance, always be the offspring of my parents; nothing can alter that. I do not, however, continue to be caused by them. I was caused by them once, although I received guidance and nurturing from them ever since. They launched a new life and set me on a trajectory (not of course referring to the swift kicks earned for certain escapades). Yet I am only partially a product of my upbringing and training. To point to an event or series of events as completely indicative of who I currently am is to overestimate the power of the past. Our self-determinative power would see to that.

Whitehead's occasions of experience act only to bring themselves to completion or satisfaction; and only perished, that is, objectified, entities can act for others. Yet act in what way? Efficient causality means exerting an influence upon, as though one object enjoys agent status over one which is merely patient. But instead of thinking of the past in terms of imposing a force, it might be better to read it as that which is drawn upon. This would turn the tables on presently forming occasions and make them active in drawing upon the past as a type of library of information, whereby the past would then be rendered passive. For Whitehead, it is not the case that only contiguous objects—those sharing a common border or boundary—can be in causal interaction, since internal relatedness gives us access to a wide-ranging past, some of which is our own. Otherwise there would be an unbridgeable gap between present and wider past such that one could not be connected directly to the other.

In Whitehead's view, the influence of the past is massive and impressive, while an incipient occasion is impressionable. This allows the process perspective to contain an element of continuity as opposed to being purely episodic. Granted that the movement can be described as quantum mechanical in character with its discontinuity, indeterminism and nonlocality, but the constant link with the past makes it relativistic or even Newtonian as well, namely continuous, deterministic and local. God's intention that the world remain orderly is played out as the past is inherited by the present by way of natural law. This makes the past a force to be reckoned with. Griffin even goes so far as to state that there is an "inrush" of the past into the present (Griffin 1988, 23). Yet to claim that the past is efficiently causal may be hyperbolic. It might be better to say, as Peirce did and as Sheldrake does, that the past is habitually reenacted in the present (although, admittedly, Sheldrake sees this as brought about by formative causation, albeit in a non-energetic way), unless (and this is where Sheldrake parts company with most others) the germ of a new form surfaces and through repetition also becomes entrenched into the present. After all, there is a certain finality about things that have slipped into the past.

Allow me to conclude these thoughts by stating that if the past fails to be causal in the way Whitehead describes, then the rest of his system is in jeopardy. There is no question that the past impinges on the present; this is not in dispute. But this might not be what Whitehead has in mind with causal efficacy. The difference might be the degree to which the past becomes an imposition, otherwise it would be less than causally efficacious. In Whiteheadian terms, God cannot coerce, but the past in fact must, otherwise a new entity could not be formed. (An interesting question would be whether initiating each process in fact constitutes coercion on God's part. Intervention as such it would not be, since God is involved in keeping the mechanism a going concern.) In exploring the Whiteheadian scheme as a suitable candidate for a new metaphysic, one must decide if one's own past has compelled in the above way. In my estimation, an active past is actually present since that is when it acts.

As I conceded earlier, these musings do not amount to a conclusive argument against the past as causal nor are they intended to be, but they do shed some light on the degree of my discontent toward this notion. If it serves only to cast doubt on a principle upon which the process framework rests, then that will suffice. Others can take it from here if they choose.

In summation, in as much as the past is effectual, it is no longer past but present. And all of the foregoing leads me to suspect that the Whiteheadian map will not take us where we want to go, but will only lead us part of the way.

Whitehead and My Discomfort Zone

The process viewpoint can be traced back to the ancient Greek philosopher Heraclitus (died after 480 B.C.E.), who understood all things to be in a state of flux and where becoming is preeminent over being. This is also indicative of Whitehead's scheme, which Griffin refers to as panexperientialism—a kind of pluralistic monism, where "there are lots of actual entities, but only one logical type" (Griffin 1989a, 25). These newly forming entities are impacted by the past, as we have just outlined. The immediate past is contiguous with the present (Griffin 1989a, 29), which means the two eras are either in contact or are next in order or time. Note that this description contains spatial as well as temporal overtones.

While space is equivalent to time according to relativity, given the appropriate conversion factor, the two can be distinguished in our experience. Yet the distinction is easier to make if past and present were regions as opposed to eras, for then the two could be compared from the same temporal vantage point. But to compare two eras while occupying one of them is more difficult than, say, assessing whether the grass is greener elsewhere. From our perspective we can legitimately compare immediate versus distant past. For Griffin, the more remote the past, the more weakly it will be apprehended by present subjects and may not even rise to consciousness, although in principle it can do so (Griffin 1989a, 29). The distant past may no longer enjoy a common border or boundary with the present, but Griffin's response is that thanks to quantum mechanics, influences can also be seen as nonlocal or noncontiguous (Griffin 1989a, 31).

Griffin takes the experience of a past event, through memory for instance, as the flipside of the statement that efficient causation is at work from past to present (Griffin 1989a, 30). For in this case, the present actually calls upon the past, which is then pressed into service. To experience, then, is to be impacted causally. If this is a correct reading of Griffin, is he overstating the case? From the process perspective, I suppose that anything can be made to appear as an experience to which we are internally related and which therefore bears a significant emotional import for us. Yet does this describe the workings of an ontological mechanism or simply the function of language? Or if our imagination shows difficulty in disentangling an occurrence from its effect on us, does that require that the two be intertwined or necessarily commingled?

This is all that I will say about the past. Now I want to discuss the prospect of divinity in a Whiteheadian strategy. His process framework is panentheistic in nature, where God contains or includes the world but extends beyond it. The world is part of God and in God. Both God and the world transcend their inherited past by virtue of their expression of self-determination. As mentioned beforehand, God presents nascent occasions with an initial aim. These ideals are embedded in objects which are then felt or apprehended by newly forming subjects as they take account of the data in their environment or past world of experience.

Whitehead spoke of our encounter with these ideals as our "experience of the deity of the universe" (Griffin 1989b, 155 n.32).

My question is whether this is the extent of our access to the divine? According to Griffin, we have a continual nonsensory exposure to deity: "In those rare moments in which this constant apprehension of the divine reality rises to consciousness, we enjoy the 'experience' of the holy" (Griffin 1989b, 66). Notice that he believes that the instances are rare. Is this a failure on the part of a God who makes him/her/itself scarce, or a failure on ours who have not as yet developed the intuitive equipment to recognize God's presence? Whatever the correct version may be, as it stands our experience of God's ideals for us amounts to little more than a inter-departmental memo or fax—a message without a messenger. In essence, is all that we ever discover of God only a paper trail?

On the same topic of initial aims, does God, on the one hand, always impart the type of ideal which exemplifies the best decision for the recipient occasion to make, or on the other, that aim most likely to take hold and have a successful appropriation? In essence, does God take calculated risks and play the odds? If the former, then God's intent is the deciding factor and God hopes for the best, come what may. The ideal is to be upheld even in the face of opposition and probable refusal. God might then elect to work with what is available and enlist those tactics which could enhance the likelihood of an unpopular aim's acceptance. If the latter, God may need to settle for second best or an intermediate step and wait for an opportune time in the next phase. A strategy might then take place in stages.

I also wish to comment on the thorny issue of God and evil. While this topic deserves a much greater and more in-depth treatment, I would be remiss if I did not at least touch on it. The ancient Greek philosophers addressed the concern of God's relationship to the world and argued that a perfect God could not be moved by what occurs in the world. Perfection for them required a static being, so God must be immutable (changeless) and impassible (unmoved). The price of keeping these attributes intact is the lack of feeling God must have for both the joys and sorrows of the world. God is invulnerable to the height and depth of the world's experience; its elation, tragedy and calamity do not touch God. This is the mindset which Christian theology has inherited.

Whitehead objected to this analysis and submitted in its place his own insight that the true nature of divine perfection includes relations to creatures and becoming enriched by their responsiveness (Birch 1998, 247). In this way both the world and God make a valuable difference to each other's development.

Whitehead's effort to deal with the problem of evil was spurred on by the death of his son in the First World War. One way to cope with the loss and come to terms with it is to propose that God can act only by persuasion and not coercion, thereby intimating that God lacks sufficient power so as to overcome evil on God's own. God requires the cooperation of consciousness-bearers, who perhaps are the ones that perpetrate the most evil, to effect this change. Within the God-world axis, the position tends to be held that it is the world-pole alone, studied by the sciences, that experiences movement. The advantage of Whitehead's approach is that God is acknowledged as a participant in the process—a "fellow-sufferer"—who also encounters evolution and can commiserate with those in pain. Yet one must ask which view is more adequate: a powerful God who engages in self-limitation, as in what is known as the kenotic model, but could do more to circumvent evil and does not; or a less potent God who holds nothing back and does everything in God's limited power to forestall it? This is a question with which many persons continue to struggle.

My own misgivings about the process framework stem largely from the concept of empowerment, or rather the lack thereof. To my thinking, there must be a force greater than simply the advertisement of ideals to enable us to live transformed lives. A turnabout of lifestyles is not easily enacted, some might even argue that it is beyond our capabilities. Such a radical change may not be something that we could manage under our own steam, and if this is the case, then some power external to us would need to inject what is lacking. Our own efforts would then need to be supplemented in order to bring to fruition the very ideals that God asks us to grasp. Nor would this be a coercive force if it operates at our invitation, upon the recognition that our attempts at transformation, being too feeble, require assistance.

Mention should also be made of one more item. I want to address the concept of societies in Whiteheadian thought. The regnant or dominant member of a society may organize the activities of the other members, but it remains a member. How does it arise to become dominant, thereby creating another level on the hierarchical scale? A dominant neuron or brain cell might coordinate the efforts of others like it, but does this mean that a mind has appeared or only a function that directs neural traffic? And how does it come about that a member is authorized to take on this responsibility and is entrusted with that charge?

There is a position known as emergentism which states that mind, for example, is an emergent property of a certain sophisticated type of neuronal activity. Like a magnetic field is generated by some body, so a mind is said to emerge from a brain. Proponents of this view have called this a "field consciousness" or "soul field," which is grounded in a physical form but is also distinct from it. The object, for instance, is localized, but gravitational or magnetic fields spread out from it "in all directions" (Hasker 1983, 73). All masses exert a gravitational force, but some objects can become magnetized. So the question boils down to two alternatives: if everything has a soul-field like all objects are gravitational, then the field is not emergent but inherent; and if some objects achieve a soul-field like some objects become magnetized, then one must explain how such a field arises from no field at all. Whitehead begins to overcome this difficulty by claiming in his metaphysic that mentality of some sort, meaning the power of self-determination which transcends the givenness of the settled past, is pervasive throughout the natural order. The point of this exercise is that if Whiteheadian dominant members emerge, then process thought is an emergentist position and must adjust itself accordingly. If not, then a proper distinction will need to be drawn.

David Bohm

If our current map is no longer suitable, then now is the appropriate time to search for a substitute. An insightful and innovative proposal is submitted by the late physicist David Bohm (1917-92), whose thought exhibits Whiteheadian parallels. Like Whitehead, Bohm rejects the assessment that reality is composed of independent fragments; and if we are taught to think in terms of such division, then this is what we will perceive (Bohm 1980, 2). Training of this sort leads to the impression that thinking corresponds to objective reality, prompting observers to accept distinctions as descriptions of the real (Bohm 1980, 3). In Bohm's diagnosis, a mechanistic philosophy is responsible for this mindset. He judges that this mentality was itself based on "fairly crude observations, which demonstrates the danger of deciding a final philosophy on the basis of any particular observations; even our present observations may be too crude for something still deeper" (Bohm 1988, 64).

Bohm shares with Whitehead the perspective of the essential nature of process (Bohm 1980, 48). Bohm employs the image of a flowing river as that which best illustrates the world's processive character. He notes that the patterns in the stream are ever-changing as "vortices, ripples, waves, splashes, etc., ... evidently have no independent existence as such. Rather, they are abstracted from the flowing movement, arising and vanishing in the total process of the flow." Such imagery drives Bohm to present a scheme with a "seamless enfolding, in which not only matter and energy but also space and time are brought together in one vast hologram of the universe" (Templeton & Hermann 1988, 126). Bohm refers to the order that science deals with as the *explicate order*—the world filled with objects. In Bohm's appraisal, there also exists an *implicate order* which he explains with the help of the three-dimensional image of a hologram, involving

> a photographic record of the interference pattern of light waves that have come off an object. The key new feature of this record is that each part contains information about the whole object (so that there is no point-to-point correspondence of object and recorded image). That is to say, the form and structure of the entire object may be said to be *enfolded* within each region of the photographic record. When one shines light on any region, this form and structure are then *unfolded*, to give a recognizable

image of the whole object once again. (Templeton & Hermann 1988, 177)

Bohm's central concern is the expression of "the unbroken wholeness of the totality of existence as an undivided flowing movement without borders" (Templeton & Hermann 1988, 172). He evaluates the implicate order as well-suited for describing reality in this fashion since in it "the totality of existence is enfolded within each region of space (and time)." This means that any "part, element, or aspect we abstract in thought, ... still enfolds the whole and is therefore intrinsically related to the totality from which it has been abstracted." Bohm thus holds to the enchantment interconnectedness view of reality as he bases his outlook on the part enfolding the whole as in a hologram. Whereas conventional science treats the part within the whole, Bohm stresses the reverse—the whole within the part. And further contrary to traditional science, the abstract world of objects in the explicate order is grounded in the implicate, making Bohm's ideas converge with Whitehead's at certain points.

Another analogy Bohm employs is the unscrambling of "a TV signal with information enfolded in an electromagnetic wave, which the TV receiver [then] unfolds as a visual image" (Barbour 1997, 177). Bohm's perspective displays a wholeness where connections can be instantaneous, nonlocal and noncausal in apparent opposition to Einstein's theories. As Barbour explains,

> Events separated in space and time are correlated because they are unfolded from the same implicate order, but there is no direct causal connection between them since one event does not itself influence another event. (It would be like two TV screens showing images of a moving object taken from different angles; the two images are correlated, but one image does not influence the other.) The theory does not violate the relativistic prohibition of signals faster than the speed of light, for there is no way to use it to send a signal from one detector to the other.

Recalling a previous theme, in 1935 Einstein teamed up with two colleagues to carry out a thought experiment about two particles moving quickly in opposite directions with a signal travelling from one to the other. The way the imagined experiment was set up made it appear as though information about the second, or receiving, particle could be obtained by the researcher in a way that would exceed the velocity of light. Prior to the time required for the signal to travel between the two particles and arrive at the second, information about the second, such as the direction of its spin, could be secured. This prompted the collaborators to conclude, contrary to Bohr, that particles possess definite properties even before they are observed, given that the speed of light is an absolute that cannot be exceeded. In technical parlance, this means that locality prevails over nonlocality, where the velocity of light is a limitation for the transmission of any causal influence between two particles. Einstein argued that there must be hidden variables in the system to account for this possession of definite properties.

Thirty years later, John Bell was able to put this thought experiment to the test, whereas before his time it was not technologically possible. Both his as well as the follow-up experiment conducted by Alain Aspect in 1983 indicate that at least one of Einstein's two assumptions—locality and definite properties—must be inaccurate, which suggests that there are no hidden variables acting on the system. Bohm agrees that there are no such variables lying undetected from our senses. Instead, what is hidden is not another variable but another order—the implicate order, which is not subject to the constraints of the manifest or explicate order, such as the speed of light. There might also be a multitude of additional orders beyond these two.

For Bohm, there are external relations in the explicate order and internal relations in the implicate (Bohm & Hiley 1993, 359). "The implicate order is general and necessary, while explicate orders are particular and contingent cases of this" (Bohm & Hiley 1993, 361). The more explicate an event the more manifest it will be and the more readily it can be captured in a laboratory setting. The more implicate an event the more subtle it will be, making it reminiscent of the quantum world (Bohm & Hiley 1993, 362). There appears to be something analogous to a semi-permeable membrane or barrier between the two (or more) orders such that essentially explicate structures of a sort might inhabit the implicate order, and intrinsically implicate structures might populate the explicate order. In addition, "the whole cannot be unfolded together in one single explicate order" (Bohm & Hiley 1993, 361).

Bohm was a grand speculator in this regard. He was struck by an hiatus in our scientific knowledge and reflected on what might fill it. The lacuna is this:

> a large crowd of people can be treated by simple statistical laws, whereas individually their behavior is immensely subtler and more complex. Similarly, large masses of matter reduce to simple Newtonian behavior whereas atoms and molecules have a more complex inner structure ... between the shortest distances now measurable in physics (10^{-16} cm.) and the shortest distance in which current notions of space-time probably have meaning (10^{-33} cm.), there is a vast range of scale in which an immense amount of yet undiscovered structure could be contained. Indeed this range is roughly equal to that which exists between our own size and that of elementary particles. (Bohm & Peat 1987, 93-4)

Bohm refers to his view as the "causal interpretation," which militates against modern physics as ordinarily understood, for it "assumes that as matter is analyzed into smaller and smaller parts, its behavior grows more elementary. In contrast, the causal interpretation suggests that nature" at these levels "may be far more subtle and strange than was previously thought" (Bohm & Peat 1987, 93). Nor are elementary particles absolutely invariant, since they "can be created, annihilated and transformed, and this indicates that not even these can be ultimate substances but, rather, that they too are relatively constant forms, abstracted from some deeper level of movement" (Bohm 1980, 49).

At the fundamental level of reality is what he calls the "quantum potential," which "is determined by the quantum wave field, or wave function" (Bohm & Peat 1987, 89). It is continuous but gives rise to actual quantum phenomena. In this way Bohm attempts to incorporate both the relativistic world of Einstein, first and foremost, as well as the quantum world of Bohr and Heisenberg. And in a manner reminiscent of Whitehead, Bohm announces, in contradistinction to contemporary physics, that "in some sense a rudimentary mind-like quality is present even at the level of particle physics, and...as we go to subtler levels, this mind-like quality becomes stronger and more developed" (Bohm & Hiley 1993, 386). Adopting Whiteheadian terminology, he adds that "[a]t each level, there will be a 'mental pole' and a 'physical pole' ... even an electron has at least a rudimentary mental pole, represented mathematically by the quantum potential." And conversely, "even subtle mental processes have a physical pole. But the deeper reality is something beyond either mind or matter, both of which are only aspects that serve as terms for analysis" (Bohm & Hiley 1993, 387).

Particles in the explicate order are actually "bundles of energy of various fields" (Weinberg 1992, 31). As Bohm and Hiley outline,

> the essential qualities of fields exist only in their movement. We propose to call this ground the holomovement. It follows that ultimately everything in the explicate order of common experience arises from the holomovement. Whatever persists with a constant form is sustained as the unfoldment of a recurrent and stable pattern which is constantly being renewed by enfoldment and dissolved by unfoldment. When the renewal ceases the form vanishes.
>
> The notion of a permanently extant entity with a given identity, whether this be a particle or anything else, is therefore at best an approximation holding only in suitable limiting cases. (Bohm & Hiley 1993, 357)

One advantage of Bohm's implicate order, on the one hand, is that it yields no gaps—it is all a seamless and unbroken whole, even though it generates, in the explicate order, that which is particulate in nature. And if one were to ask, concerning this ballet of enfoldment and unfolding, who or what is arranging the choreography and how?, one commentator responds in this way: "It is the holomovement itself that exclusively determines what is to be unfolded. The world is hoist on the holomovement.... Unfolding is the sovereign province of the holomovement" (Globus

1987, 376). One disadvantage, on the other hand, is that there may be the introduction of a needless complication, specifically a multiplicity of implicate orders whenever they are theoretically required. As Bohm and Peat declare, "for the second implicate order, the third is a more subtle mental side and so on" (Bohm & Peat 1987, 211). Facing such a cumbersome possibility, Bohm is quick to admit that his system is not the last word. It too has limitations which might need to give way to, or become incorporated into, "some yet more comprehensive idea" (Bohm & Hiley 1993, 390). In a similar vein, these authors do not assert finality for their ideas but make them "points of departure for exploration."

In answer to the question as to how many categories of things there are in the universe, Bohm would likely have said one—the essence of a thing remains the same though its form is changed, depending upon the order in which it happens to take up residence at any given time. There is one category of things with a succession of orders to accommodate them, and this makes his scheme monistic. To his thinking, "[o]f course this does not imply that 'consciousness' can be imputed to electrons or to other such 'particles.' This arises only at much deeper levels of the generative order" (Bohm & Peat 1987, 211). Yet "there is no absolutely sharp 'cut' or break between consciousness, life and matter, whether animate or inanimate."

In my view, Bohm's attempt is largely successful in overcoming dualism. An explicate entity is a transitory product of the implicate. This is not a hierarchy, where one manifestation is higher or lower than another. They are merely different states of the same reality, although deeper levels or orders come into play. An electron, for instance, is at one point an unfolded, observable, manifest "wavicle" and in another phase it is an enfolded, unobservable, unmanifest "denizen of the deep." Each is potentially in the form of the other; one is formed and the other is unformed or currently without form. Neither is any more or less real than the other.

Perhaps this is a map worth consulting. The question is: can it stand alone or does it need some assistance?

The Combination Approach: Syncretic or Synergetic?

> *The physicist David Bohm says "Thinking within a fixed circle of ideas tends to restrict the questions to a limited field. And, if one's questions stay in a limited field, so also do the answers."* (Watson 1980, 205)

When I patronize an establishment for the consumption of quick comestibles (a restaurant of the fast food variety), invariably I am asked the question, "Would you like fries with that?" At which point my thrifty constitution is engaged and I carry out some hasty arithmetic on my feet so as to determine if the "combo meal," including as it does a beverage, would be a better value than just the straight sandwich and fries. Deciding that the solids would likely make me work up a thirst, and ever vigilant about the best deal, I opt for the combo, confident that I have made the correct consumer choice. I then survey all the victuals on my tray and sit down to a fine repast.

When we consider some of the maps examined thus far in our analysis, we are left to wonder if there is a combo meal to be had even here. Could multiple maps be superimposed without too much difficulty? And would this produce an irreconcilable blend of distinctives as in an attempted syncretism, or a cooperation as in a synergism where the various facets work better together than individually?

Maybe the concept of fields holds the most promise. Let us ruminate over it in the following way. According to conventional scientific wisdom, both space and time were born with the beginning of the universe—space expanding and time "elongating." The material of the early universe and its associated fields are equiprimordial in the sense that a highly charged field accompanied a highly contorted spacetime as it unfolded. Fields of the type that Bohm envisages are in the business of yielding particles and perhaps even directing their traffic. Particles become local condensations of energy which behave materially. If God were to be linked to a field, then God would enjoy intimate interconnections with the world in such a way that light could be shed on the

Christian doctrine of the Incarnation. God as in or behind or coterminous with a field, whatever the appropriate preposition might be, would be involved in the careers of all cosmic particles, thereby rendering the idea of intervention obsolete. Involvement in the world would be the ordinary state of affairs, implying that an incarnation would lose its status as completely unique. All particles would begin and end in a universal field that could be regarded as an extension of God. This, of course, would be maximally controversial to traditionalists, but it demonstrates the effect which scientific and philosophical speculation can have on religion and theology.

The next issue is a more Whiteheadian one: the degree of motion that God imparts to the particles versus the amount of self-motivation which the particles have at their command or have been delegated. A particle is never bereft of divine presence if not outright influence, and God is customarily not without reference to particulate expressions. Contrary to Bohm, sufficient self-movement could enable particles to "kick" each other directly without the signal needing to be rerouted through God's field. Internal relatedness could enable such influence.

A situation such as this would also have Sheldrakeian contours, for some particles would be mandated with constancy, thereby forging an ever-deepening groove in the background field with each successive generation. This would compel the trajectory of particles essentially to replay the forms and careers of their predecessors. Other particles would be commissioned with the task of embodying a greater creative component to their life histories and casting a wider net into the risky probabilities and indeterminacies which confront them, so that they, in so far as they have the capacity, may make the selections at their disposal. The initial or primary stage is ordinarily understood as the inorganic world studied by the physical sciences. The second stage of particulate arrangements is usually investigated by the natural or life sciences. Then at the tertiary stage, certain collections of particles are empowered with the capability of reflecting upon the nature of reality and hence demonstrating the kind of behavior treated by the psychological and social sciences. And with repeated attempts, the field is likely to yield more of the same forms and behavior.

One of the key issues to be addressed here would be the conferring of potentialities upon particles on the part of the God of the field. In so far as this is what actually occurs, then nature contains properties which the scrutiny of the physical sciences cannot exhaust. If they are latent, then they could be activated given the proper circumstances—a function exercised at the discretion of the field-God. Without getting too far ahead of ourselves, this may effectively bridge the gap between the organic and inorganic worlds, mind and body/brain, and soul and spirit. Each level would be manifested in conjunction with the cosmic field that generates it. Yet a question surfaces as to whether fields bear the wherewithal to confer such qualitative, as opposed to purely quantitative, power to certain of its regions, for the issue is not one of the degree of power but its type. Were it simply a matter of structure, then quantitative concerns would suffice; but the emergence of different kinds of operation suggest that different properties are at work. A talent or skill is not normally associated with or affected by the injection of additional power.

In my perception, this indicates a fragility within the Whiteheadian distinctive of the emergence of an organizing center, or what he calls an "all-inclusive" or "unifying" subjectivity. This marks the mysterious dominant member which coordinates the other members to produce consciousness as the distinguishing feature of some societies and self-consciousness in humans. Where it originates and how it arises if it is not found elsewhere in the hierarchical scale is, at least for myself, unclear.

Perhaps Whiteheadians, as intimated above, might claim that it is latent in precursor structures awaiting the full flowering of forms bearing greater sophistication. It is precisely this sophistication, however, that is at issue, for does the arrangement propel the sophistication, in turn igniting or triggering the dominant member into action, or does the organizing center make this arrangement sophisticated? In the alternate format that I suggest, as the field bears potentials which actualize sometimes this and other times that fundamental particle, each intimately interconnected, at least initially, so too emerges from the field a gradation of abilities to organize the particles within its regime. As nature can boast potentialities not always readily apparent through standard experimental means, so does the field marshall possibilities that might otherwise be unforeseen.

There is no emptiness to space for Bohm. By virtue of fields, or in Bohm's case the implicate order field which he names the quantum potential, the universe is pregnant with particles. The field is required for particles to be born, but there is no field without a physical counterpart. It is characteristic for a field to generate particles, and fields are usually not dormant in doing so. And Sheldrake's is the only field other than Bohm's whose "influence does not diminish with distance" (Talbot 1992, 39-40). All others, including the past for Whitehead, trail off or fade with time or space. Fields produce particles and bodies or objects produce fields, and each exists in conjunction with the other. This lends credence to the process notion of an eternal universe—one must always have existed, for God must always be embodied with a world.

Yet does there need to be some type of involvement without which particles will not be created? Must there be something like local concentrations or condensations of energy to allow the production to commence? Some researchers might claim that the situation may be like crystals which begin to take shape around an impurity in a supersaturated solution. But does this adequately describe the cosmos? Bohm declares that "if the implicate order [were] totally formless, then form is an illusion." He further "propose[s] that each moment of time is a projection from the total implicate order" (Bohm 1986, 189). In Whiteheadian terms, if there exists an opportunity for creativity to be instantiated in an event that is becoming concrete, then the field would likely seek such an expression. Whitehead's God could also be analogous to the implicate order which reabsorbs particles, or to frame it using process terminology, God takes up particles into God's own becoming once they complete their own round of becoming. Particles would then shape the field, in addition to being shaped by it—a situation that would be agreeable to Sheldrake, for whom fields evolve, but perhaps less so for Bohm.

Such are the contributions of and by Bohm, Whitehead and Sheldrake to this speculative amalgamation. The question remains, though, as to who or what is responsible for the field or order?

The Forces of Nature

There seem to be four options with respect to the question about how many classes of things reside in the universe. If the answer is one, then we have a monism with two alternatives: if there is matter only, then materialism rules; if there is mind only and material reality is an illusion, then either the idealism of Bishop George Berkeley prevails or the pantheism of Spinoza (or a school of thought from certain Oriental philosophies). If the answer is two, then the possibilities are these: mind and matter are treated either as ontologically separate, in which case there is a Cartesian dualism such as the transcendental supernaturalism found in certain Christian circles; or they work in conjunction meaning a Whiteheadian or Bohmian holism would apply.

Advancements in the issue of divine activity and the God-world relation would be made if clarity could be obtained as to which natural phenomena receive God's ministrations, along with the additional concern of whether this involvement is pervasive or localized, continuous or intermittent, versus which phenomena are on their own and accomplish what they do unaided. Or perhaps, as with Whitehead, there is never an occasion of experience that goes without divine attention and input. A further look at the operations of nature might help us here.

Barbour, for one, adduces the inherent abilities of the organic world in stating that not all combinations of organic molecules have an "equal probability, for there are built-in affinities and bonding preferences" (Barbour 1998, 424). Some chance arrangements reveal a greater stability than others, and what the organic world consistently displays is "a capacity for self-organization and complexity because of structural constraints and potentialities." Moreover, pattern becomes an important consideration in these deliberations, for it "develop[s] in the whole without prior specification of the parts" (Barbour 1998, 428). Barbour understands this "readjustment of the parts" to be the work "of top-down causation" and cites the example of an infant to this end: "the brain of a baby is not finished or 'hard-wired' at birth. The natural pathways are developed in interaction with the environment and are altered by the baby's experiences." Hence the child plays a role in its own neural formation. And as for the mind-body/brain problem, Barbour attests that mentality and physicality must overlap at some point, for no amount of one can generate that which resembles the other (Barbour 1998, 441). Yet the aforementioned concept of pattern might be suggestive here, enabling matter to exhibit properties not yet fully appreciated. If so, less appeal to divine activity of an interventionist stripe need be made. If the pattern of the whole can supply what the parts alone lack, then such an investigation should first be exhausted so as not to render hasty theological or metaphysical judgments.

The category of mind is a particularly fertile one in reference to the involvement of deity. Recall that in the quantum realm indeterminacy means that outcomes are not predetermined. If a system is in a superposition of states, interference with it causes one of the multiple states to be assumed. If the cosmos is managed by a deistic divinity, so the argument runs, then God's preparation of the initial conditions of the universe is insufficient to bring about God's purposes if the system requires continual determination (Clayton 1997, 208). This of course would be true unless God has no vested interests in definite results. Perhaps God appreciates the element of surprise both when giving and receiving it. In application to the matter of mind, divine "guidance at the quantum level could produce an otherwise unlikely brain state (which the individual may experience as a particular thought) or a gene mutation that would have specific effects" (Clayton 1997, 214). Supposedly this would be the case whether one is awake or asleep. The point is that indeterminacies incessantly become determined through whatever means. The question remains as to whether these means are fully natural, with the parenthetical note that nature may contain more in its arsenal than currently recognized, or if they have some divine component (and indeed if the divine has a natural aspect). Besides, the postulation of other orders, as per Bohm, sheds a different light on the issue of whether another order is responsible for the very quantum phenomena (and their outcomes?) to begin with.

There is still another spin that can be placed on nature and its properties and this comes to us from an unscientific quarter. As Peat discloses,

> traditional Native ideas of harmony and balance indicate that if order is created in the laboratory, then disorder must be created somewhere else. How ... does science take responsibility for this? There was also concern at the way physicists learned about elementary matter—by colliding particles together in a particularly violent way. (Peat 1997, 315)

Such a methodology is opposed to the Native outlook of "entering into alliances with [nature's] energies" (Peat 1997, 315). Native spirituality appears to be sympathetic to a "process-based worldview" and sees nature as personal, referring to it as "her." This approach would not endorse the standard procedure of "learn[ing] about the internal structures of elementary particles by shooting them at each other, then observing how they scatter" (Peat 1997, 45). And the sensitivity toward nature also extends to us, since we are part of it. Subsequent to our death and the decomposition of our bodies, their "constituents begin to disperse through the earth and its atmosphere, often becoming part of other human beings" (Peacocke 1998, 376 n.52). Dust returns to dust but it enjoys quite a ride in the interim.

If each part of us carries with it a minuscule reflection of who we are, including the countless times it has already been recycled prior to reaching and becoming us, then I imagine that this would be significant if we were to receive such resources from a Mother Teresa or a Jack the Ripper. We lose our identity upon death, but evidently, as Lyall Watson insists, "We carry a reminder of who we are in every body cell. A sense of identity is basic to all life, even at the cellular level" (Watson 1995, 43). Powers of identification are called upon for cell recognition, cohesion and one's immunological system. The ability to discern what is and what is not us is vital to our survival and this requires cooperation—a team effort among the cellular and perhaps even molecular components.

Hence nature and its material constituents might be endowed with greater capacities than we reckoned with and gave it credit for. Or the divinity may use them as instruments to effect the changes that comport with the divine approbation. Once again, the resolution depends on the characteristics of this endowment.

CHAPTER SEVEN

Which Way to Turn?

HAVE WE COMPLETED OUR TASK? By no means. What then is left and remains to be done? I would like to propose a rationale as well as an investigation that might take experimentation in a slightly different direction than usual, perhaps even tangential to traditional approaches, and that could be considered out of the ordinary. But before I do so, I wish to offer some additional cautionary remarks about the ground that we have already covered as we proceed on our trek, much as a caregiver might say to a child as it makes its way to a friend's home: "Look both ways before you cross the street."

On the topic of the physical sciences, it would serve us well to be reminded that we require physics in order to study physics. The book of nature is opened to us through natural means and this presents us with a restriction. We are constrained in what we can know by the parameters within which we work. The smaller an object, for example, the less likely we are to have direct epistemological access to it. Microscopes can take us only so far, since "light is far too gross a tool to probe such a tiny entity as an atom." As Peat informs us, "The wavelength of [visible] light—the distance between one peak and another—is vastly greater than the dimensions of an atom. Light simply does not 'see' atoms" (Peat 1997, 44-5). A related difficulty is, of course, the measurement problem in quantum mechanics, namely how an experimental set-up can "yield exact measurements on a quantum system if it too is composed of elementary particles obeying the indeterminacy principle?" (Russell 1998, 211), as well as the other rules of the quantum game. The new physics prompts Russell to render the following judgment: "classical physics is in principle false; as a fundamental perspective, its view of nature and its explanations of the world are wrong" (Russell 1998, 217). This would include, for instance, the non-additive nature of the velocity of light—how, say, adding one half of the speed of light to another half still does not enable us to reach light speed.

Turning from the micro- to the macroscale, there is a cosmological model known as the many-worlds theory which states that quantum indeterminacies become determinate individually but not singly. It is believed that the entire multiplicity of a superposition of states becomes expressed rather than merely one state, and this occurs in different universes. We, unfortunately, have access only to our own, so we can never know how the other states are played out. But played out they are. Davies notes that if the intent here is to circumvent the need for an appeal to divinity (as Fred Hoyle attempted in his steady-state model of the universe), then this is hardly superior to a theistic position:

> Invoking an infinity of unseen (and perhaps unseeable) universes just to explain the one we do see is the antithesis of Occam's razor [the rule of parsimony], and is fundamentally unscientific. In any case, it is scarcely more plausible than a single unseen God. (Davies 1998, 156)

And on the theme of the anthropic principle, Davies declares that the presence of life and consciousness in the cosmos, as yet detectable only on our planet, "imposes rather stringent restrictions on the values of the fundamental constants of nature and on the cosmological initial conditions" (Davies 1998, 157). While this is undoubtedly true, the question which I wish to pose is, even if design implies a designer, whether intricacy of form automatically entails design? As with chaotic systems, what we perceive as complex might actually lack remarkable beginnings.

Moving from the physical to the natural or life sciences, there is a point where molecules are promoted from the inorganic to the organic level and this has to do with the bonding affinities of the carbon atom. What is more, an atom of iron in a rock sample is qualitatively different, for thinkers like Whitehead, than one in a molecule of hemoglobin by virtue of its internal relatedness to the pattern of atoms in the surrounding structure. Nevertheless, as Randall urges, "it is a very far cry indeed from the formation of giant molecules" even if they are biomacromolecular, "to the formation of self-replicating units upon which [natural] selection can act" (Randall 1977, 203). Evolution may occur in physics and chemistry as it does in biology, but it might not justifiably be described as Darwinian. And regarding the explanatory power of genetics, while it is accurate to state that "some genes are turned on and off at just the right moments and some work in conjunction with others in just the right amounts ... the information for the execution of these tasks is not to be found in the genome itself" (Stoeger 1998, 178). This reinforces Sheldrake's claim as to which specific "time[s] genes are engaged and which ones contribute to an arm or a leg are also not to be found in genetic instructions," although a solution to this problem is not necessarily out of reach.

Continuing on the topic of Sheldrake, he finds science in its use of terms such as genetic "information," "instructions" and "programs" to hypostatize the natural world in a way that is unwarranted for a purely materialistic mindset. Matter does not instruct, only subjects can, and if science is comfortable employing terminology that betrays interiority, then it needs to modify its outlook accordingly to one expanding beyond external relations alone. Moreover, in speaking sympathetically of Sheldrake, Watson takes science as overextending the explanatory power attributed to the category of instinct. Watson gives several pertinent examples and it seems appropriate to quote him at length:

> [some] animals us[e] navigational skills with little or no apparent training. Pigeons find their way with the help of familiar landmarks, subsonic sound patterns associated with the passage of wind through certain mountains, the height and position of the sun in relation to an internal clock, the patterns of the stars, and the lines of the earth's magnetic field. [But] certain species, such as the European cuckoo ... are able to make these voyages with-

out instruction, following parents they have never known thousands of kilometers south to the traditional winter feeding grounds in Africa. (Watson 1992, 61)

Additionally, monarch butterflies

> migrate each winter from Canada down through the United States to the highlands of Mexico and return along the same routes the following spring. The boldly patterned hordes fly over 100 kilometers a day, stopping each night to rest in huge brown clusters among the branches of particular trees—the same trees every year.... And the monarchs make this annual migration, and unfailingly pick out the traditional motel trees, despite the fact that the butterflies traveling in any year are first- or even second-generation offspring of those that covered the course in the opposite direction the previous season.
>
> All these patterns may be determined by genetic programs—the new all-purpose explanation for strange behavior. But there is no known gene, no established physical force, no electromagnetic link that can act in this way to guide a cat or pigeon to a totally unfamiliar place. (Watson 1992, 63, 65-71)

Watson then goes on to focus on anthropological ramifications of genetic theory and comments on the consequences of ideation: "In the course of human evolution, a change of mind, a new idea, can have as much survival value and adaptive significance as the mutation of a gene" (Watson 1992, 10). Culture is also heritable, as we have already noted, and in Sheldrake's view the greater the degree of similarity and frequency between current and past cultural practices, "the stronger the resonant connection" between them, and the more likely that they will be repeated (Watson 1995, 237). Yet the onset of ideas themselves is inexplicable, especially for completely new ones. Instead, the latter call for "a sort of cultural mutation, with someone like Darwin or Sheldrake acting as the mutagenic agent." Cultural transmission of information, often through education, makes it a rival pattern of inheritance to its genetic counterpart. Culture, of course, can pass not only from parent to offspring but also laterally from teacher to uninitiated. Compounding its divergence from standard natural selection is the fact that the Darwinian approach neither takes into account any long-term effects, nor does it plan for the future (Watson 1995, 238).

Next, let us consider a few points in line with the work of Becker. In the electromagnetic spectrum of radiation, the range of visible light for us is demarcated by less than one order of magnitude, between infrared and ultraviolet, meaning that there is a very narrow bandwidth of human sight. In comparison, our auditory capacity extends through the range of approximately 20 to 20,000 Hz (often depending on our age). This covers three orders of magnitude or roughly two hundred times our ability to see. This implies that we are very much in the dark about what goes on in the world (and does it also mean that we are built to listen more?). Hence there are a multitude of signals that are hidden from our senses because we do not possess the equipment or faculties to detect them. Yet they may impact us despite our lack of awareness; and if not innocuous, the effects can be either adverse or beneficial.

Even though these signals are generally beyond human perception, everyone is vulnerable to them, but some persons more so than others. Certain individuals might be sensitive or prone to their effects. There are, however, specific rhythms and cycles that all organisms respond to, regardless if they can consciously identify them. Not least among these is the circadian patterns of organisms, differentiating those which are diurnal from those which are nocturnal. The reason for this sensitivity is that creatures are electromagnetic systems in an electromagnetic environment from which they cannot extricate themselves. This further suggests that the contents of the universe are linked or interconnected:

> An event originating from a distant quasar can be connected to an event in our brains. Ours is not a universe of isolated parts; we are made of the same raw material as the rest of it, and we respond to the same forces that drive and shape it all. Star and skin are in regular contact. (Playfair & Hill 1979, 355)

It is interesting to note that the author to whom we devoted nearly the least amount of space in the foregoing, namely Becker, has become the pivotal point upon which our discussion hinges ("the least shall be the greatest"). Paul Devereux and colleagues expand on Becker's work by taking his analysis one step further. They elaborate on their theme in this way:

> the human brain contains electrical frequencies that relate to those found naturally in the energy fields of the planet itself. Some researchers are convinced that this came about because life evolved within the influence of these fields, and the brain rhythms became entrained by them ... planetary pulsations [then] began to "drive" the electrical activity of the brain.
>
> [Thus] in certain mental states our brain waves are resonating with the rhythms of Earth. We can attune ourselves to the planet. (Devereux, Steele & Kubrin 1989, 77-8)

Additionally,

> living organisms are able to sense or respond to all the natural fields in which life on Earth has developed, even if many of these sensitivities are below the threshold of conscious awareness, or have changed or even atrophied as environmental circumstances have altered over the ages. Orthodox science tends always to underestimate the sensitivity of living organisms to weak energies. (Devereux, Steele & Kubrin 1989, 116)

Divine activity then could penetrate the world by overlapping with it. In essence, we are affected by the world because the divinity has a hand in it. There is thus a reason that we call our world Mother Earth—our planetary parent brings us forth, thereby placing God in the role of midwife. A strategy such as this, however, for better or for worse, might be looked upon as too coercive for Whiteheadians, unless the activity is limited to coaching and comforting.

Armed with the foregoing, I propose the following road map as a springboard for further research.

When in Doubt, Venture Out

Among the characteristics that humans share with some other life forms, two stand out in my mind as prompting extended investigation. In the first place, whatever it is about *Homo sapiens* that enables them to enjoy self-consciousness, they are the fruits of the planet, the dust of the earth. In the second, individuals are composed of other individuals. Humans are composites. They have crossed the line from uni- to multicellular organisms, originally beginning with cells that formed colonies and later with the union of two gametes forming a zygote, which is soon to divide into an embryo. In the first instance, people are earth-bound space-travellers. They are born of the earth; the planet becomes their parent. And in the second, persons are associations, and for the collective to operate smoothly there must be cooperation within the membership. Both the earth and humans are the many which, in turn, have given rise to many.

From my perspective, the above points generate two lines of inquiry. First, if our origin is at least partially planetary, then we inevitably take some of our cues from earth. However subconscious it might be, we inescapably resemble the planet to some degree. The most likely aspect of common ground between us and earth, people and planet, is in the cyclic patterns each displays. We exhibit biorhythms, and the earth georhythms. If the two get out of synch, even extremely low frequency (ELF) level differentials can cause a disruption in the proper functioning of organisms. Brain wave patterns in the alpha range (8-13 Hz) resonate with those of earth (10-15 Hz). Small changes in energy have measurable physiological effects.

Second, each cell in the human body is assigned specific tasks to perform, but the basic pattern remains—cells work in conjunction with others to produce an overall optimal effect of stasis at the organismic level. Brains are composed of cells, yet memory, for example, is not necessarily confined to the human head merely by virtue of the fact that the head is home to certain types of cells. If memory has a biological basis, this basis, to say nothing of its range of influence, may be wider than the neurons of the brain. Other organs might boast a similar faculty.

These statements prompt me to submit two complementary research proposals. Both address awareness in general and perception in particular. In treating the human-earth relationship it has been discovered that adjustment to the frequencies emitted by earth's electromagnetic field, sometimes involving geographical sites where sensitivities are more prone to become manifest, enables some individuals to recount experiences of hypnagogic effects such as vivid images, which could be interpreted as mystical. I am inclined to categorize descriptions of this sort not as extrasensory but supersensory, meaning elevated perception. Examination of persons with such heightened abilities, together with the geographical sites that might promote them, will afford us insights into our relationship to the planet, each other, and may also broaden our understanding of the boundaries of perception.

As for networks of memory storage and retrieval, a study could be made of organ donors and their recipients in an attempt to uncover any modifications in the latter's sentiments and behavioral characteristics. Some organ recipients have described alteration of interest, sometimes leading to changes in practice, such as in dietary habits, subsequent to organ transplant. Upon further investigation, it was disclosed that the change resembled activity engaged in or preferences chosen by the donor. The memory, one could argue, of the donor's personal predilections is communicated to the recipient through the transplanted organ. Such a study could be conducted through questionnaires drawn up by neurobiologists and distributed by health professionals, especially when access to medical records is restricted.

Analyses of this type would inform us about patterns that can be imprinted onto the parts of whole organisms beyond the brain and which in turn can be tapped into, even in a new setting. Potential conclusions to be drawn include memory as a capacity more field-like than substance-like and spread out over a region wider than not only specific brain centers but also the grey matter itself. This would make memory similar to how the electron is understood in the quantum interpretation. Bohm's holographic image is also suggestive here, as in fact neurosurgeon Karl Pribram unearthed prior to Bohm, concerning memory being often distributed throughout the brain and not confined to localized sites (Talbot 1992, 13-7). Yet our study would take the investigation even beyond the borders of the central nervous system.

Both of the above analyses would ultimately address the concept of fields—those of earth and those of humans, with the understanding that the two are connected. Persons could then be interpreted as more earth-like than previously suspected, and memory less brain-like.

On a Personal Note

I wanted to end with a few reflections on where this leaves God and the Bible. Over the years I have come to recognize the agenda-boundedness of the Judeo-Christian scriptural witness. This has led me to suspect that it cannot be counted on as a message automatically fostering an accurate image of the deity. Instead, it submits a rendition of divinity having its origins in human construction. No doubt the authors drew inspiration from the object of their faith; nor does this mitigate the value of the biblical writings, which God is still pleased to utilize as a source of wisdom, for the present-day. Yet the timelessness of the instruction might be overstated by those with vested interests in it. The books and letters of the Bible were written for specific purposes to specific audiences. The least we can say, for instance, about Paul's numerous requests for his readers and listeners to pray for him in his ministry (Ephesians 6: 19-20 and elsewhere) and his word to Timothy to bring Paul's cloak and scrolls, especially the parchments, to him when he comes (II Timothy 4:13), is that they are not meant to be carried out in our contemporary context.

For me, the God which the Bible portrays continues to this day, but in the culture and spacetime in which we find ourselves, our grasp of God's self-revelation is modified, as God would be, say, if the Whiteheadians had their way. We can even notice this in the development of the scriptures as the accounts of God's relation to humans proceeds from one era to the next. One may even be inclined to suggest that this signifies a didactical progression, where God is depicted as one who operates in accordance with educational stages, reminiscent of Lawrence Kohlberg, regarding God's responses to the children of Israel. If the accounts are reliable, at one time in the infancy of the Hebrews, God punishes by death outright in retaliation for the rebellion spearheaded by Korah and his band of insurrectionists (Numbers 16). At a subsequent time, death is exacted upon individuals instead of a group, such as the disobedient (read idolatrous) monarch Jehoram (II Chronicles 21). And later still, God imparts correction on those who are not incorrigible, like the Apostle Peter after his threefold denial of Jesus (Matthew 26:69-75 and parallel passages Mark 14:66-72 and Luke 22:56-62).

There are exceptions to this chronology (notably the episode involving Ananias and Sapphira in Acts 5:1-11), but the general trend appears to describe a movement from death to groups, to death to individuals (in the spirit of Ezekiel 18), to thorns in the flesh (Paul in II Corinthians 12:7-10), to stern warnings (Revelation 2:4-5), all in an effort to teach the people of God the ways of God. Perhaps humanity is now in a phase when God renders to us adult education, revealing that not only God's interaction with humans but also God's own becoming depends on and is altered by what has gone on before. Both seem to undergo evolution. As in a healthy nuptial arrangement, both sides develop in their capacity to relate and respond to each other. This by no means is intended to imply that we in our present setting get it right more often, but that our growth in community with each other and with God calls for different measures of upbringing in succeeding eras. Whereas at one point God was like a parent, now God is perhaps more like a partner (albeit a senior one).

In a similar vein, there are times when old paradigms no longer suffice and more progress would be made if they were improved upon. In such "revolutions," what is initially met with resistance may eventually constitute a stage of new discoveries which in turn open up new vistas that might not have been imagined under the old regime. Psychoanalysis, for example, was the subject of derision until it became an established field in its own right, spawning the discipline of psychology proper. Without them our view of the human would be impoverished. And despite initial reservations, a world framed without reference to relativity and quantum theories would yield a perspective on reality that is retrogressive because it is incomplete. The Newtonian mechanistic view is applicable to our everyday existence but is inadequate for explaining the structure of the world at scales far above and below our own. Yet as more mystery is unravelled, more is encountered. Science does not address "why" questions, nor can it (although questions such as "why does the moon sometimes appear red or blue?" abound); but that is just as well, since it has enough trouble with "what" and "how," such as what happened at $t=0$? and how did life arise?

In all of God's dealings with persons, we are dependent upon the type of revelatory messages which we as humans can process. These must be received in ways that we can capture, digest and respond to, otherwise, contrary to Karl Barth, the divine dispatcher need not bother. At the same time, it would be beneficial to be open to resources that have remained largely untapped. As we investigate with openness, we may receive what nature has to convey. Perhaps there is a vision which mainstream thinking has been reluctant to entertain but which might offer an abundance of insights. This could surface (and this is the reason that I broached the topic of psychology) in the area commonly known as parapsychology. Whitehead himself makes a few obscure references to telepathy, for instance, and could incorporate it into his system with his treatment of internal relatedness. Process thought intersects with parapsychology, as the title of a recent volume by Griffin indicates,[12] making it a field that would hold clues to what is essentially human and perhaps even divinely-driven. Sheldrake also contributes two current works on sixth and even seventh senses.[13]

There are pitfalls aplenty in these endeavors, to be sure, but if God has had a hand in shaping reality this way and is eager that people stumble upon it, then we had best not dismiss it out of hand, but rather get on with the task of examining what may have been bestowed upon us. Each of the authors in our study has something of value to offer, but the question is what and how much? And if one wishes to opt for a "package deal," then the attempt to juggle multiple perspectives might result in our barking up the wrong metaphysic. Venturing out is potentially hazardous, but the possible rewards are equally as significant. Any undertaking has its risks. This one might well be worth taking.

One more thing before I close.

An Initial Final Word

Maps are odd things. Sometimes you can't get them to fold back up properly after reading them. And they don't always furnish you with all the helpful information that you could use. They do not tell you, for example, which roads are being resurfaced so that taking them means that delays may be expected. Nor do they caution you as to which routes are dotted with potholes that could inflict damage on vehicles. Hence maps are neither up to date nor yield extensive detail.

Besides, when one is having difficulty reaching a destination, a sinking feeling can come upon one that the proper map is not being consulted. One option at that point is to ask for directions, but this of course requires the presence of at least one passerby to give them. With the rarefied air that we have been breathing lately, there may be no one else around at these lofty heights. If we are still within the gravitational pull of familiar surroundings, then this would be an opportune moment to collect ourselves and check our compass, lest we lose our bearings.

This might be the last word of the present volume, outside of the appendices, but I trust that it is only the beginning of fruitful research into what science has traditionally deemed as forbidden territory. Perhaps cartographic skills are in order as we proceed, so that we can draw up a new map while on route.

APPENDIX ONE

God by Any Other Name

THE NICE THING ABOUT APPENDICES is that they are unnecessary. This one is no exception. And before I get around to what I really want to say, I need to preface it with the following.

"When I use a word, ... it means just what I choose it to mean—neither more nor less." Humpty Dumpty should know, after all he pays the terms he employs (literally) in accordance with the amount of work he asks them to carry out. If a particular word is overextended—if it is called upon to refer to something beyond its conventional borders—then it can expect to be paid extra for this service. When asked by Alice as to whether words can be controlled in this fashion, Humpty Dumpty replies with an alternate question: "which is the master?"[14]

Who or what indeed has the final authority when it comes to the parameters within which a word may operate? Is it conferred upon them by their users, terms thus becoming tradition-bound, or can these same users modify a word to suit new needs? This prompts the wider issue of the nature of language itself, specifically the extent to which a language is fixed. New terms are being added to a language each generation and others undergo renovation. Lewis Carroll, for instance, introduced the words "chortled" and "galumphing" into English vocabulary.[15] "Facsimile" now has the additional meaning of a text that is sent electronically, and "nerd" has also made its way into dictionaries, having the meaning of a person who lacks style and class, in essence one who is not "cool." Hence language is not fixed but fluid, it evolves. This makes dictionaries snapshots of the way a language functions, which means these volumes are useful only for the era in which they are produced. They are historical documents. They describe the way a language works at the time of printing, they do not prescribe the manner in which a language should operate long-term.

Returning to Humpty Dumpty's question as to who the master is to be—words or those who use them. An analogy may be helpful here. Perhaps the word-user relation is similar to the situation surrounding elected officials in a democracy. The people within such a political system have the power to elect into office a person of their own choosing. They are the masters as to which individuals will hold which offices. Once the votes have been cast, however, and the successful candidate sworn in, the people in the democracy come under the authority of this official, in whatever jurisdiction applies to his or her portfolio and for the length of his or her term in office. Once he or she has outlived his or her usefulness, the process can begin again.

Similarly, meanings are conferred upon words by convention so that communication might occur without confusion. If confusion does arise, then either additional terms or further shades of meaning for currently existing words are developed, again by convention, to avert subsequent difficulties. Once this mutual agreement is in place—once, that is, the users as masters have acted—words become the masters and can be used with confidence that they will have only those intended meanings. The restriction or limitation then rests upon us. Until such time, of course, as they are declared unfit to rule in the standard way. At that point, the users become the masters once again and either elect a new term or a new task for the same word to perform. Both words and their users appear to take their turn being masters.

With this in mind, how does the term "God" fare? When I have been asked to speak at certain gatherings, sometimes I am asked the question as to whether there is not an alternate word or name that can be substituted for "God." After all, if the Whiteheadians are correct and God evolves, then perhaps the name can as well. I am finally delivering on this request, or at least making an attempt. To this end I offer three suggestions.

There are few terms in the English language that evoke greater discontent than that of "God." But the idea of God seems to be not so much the subject of attack as it is devoid of meaning. People object to the word because of its overuse—it has denoted so many things to so many people that it has lost all meaning. Despite the term's fatigue, some attempts have been made to salvage its content if not its name. Others felt that the word could be retained while its meaning broadened to include an immanence in the world; nature thereby coming within the sphere of divine influence. A related project sought to drive God into the metaphysical depths by portraying God as the "Ground of Being" or even Being itself. Yet for those for whom becoming is at least as important as being, disapproval continues to be voiced. Their "ultimate concern" was not God's existence, the term having an actual referent for them, but activity.

There is no ready-made solution to these problems that is agreeable to all. One band-aid remedy appears to have been adopted by some writers who have either replaced one vowel for another, namely the "o" in God with an "i" (Gid), or omitted the vowel altogether and substituted a hyphen for it (G-d). My first proposal is a metaphorical alternative. Instead of "God," one could use the term "Guide." This may be appropriate for the following reasons. First, the name reads and sounds like the one we have grown accustomed to but have become uncomfortable with. We replace a short "o" sound with a long "i." The advantage of this approach is that it is not far removed from the traditional word; essentially changing vowel sounds is not a large departure from the old one.

Second, the term connotes an action that is being carried out—a directing of those open to the Guide's leading. Some persons, though, might object that this is not their experience of God. For them, God does not guide, nor is this even desirable to, say, a Whiteheadian. Third, the concept implies a conscious personality either involved in the activity or lying behind it. Think perhaps in terms of a tour guide who walks people through unfamiliar terrain and points out signposts along the way. This might, however, still be too coercive for process thinkers, yet at least we can surmise that those who are on the tour have chosen to be there and to this extent God acts persuasively. They might select a name like "Presenter" or "Transmitter-Receiver" (TR) instead.

God has been called by many names, depending upon which of God's multifaceted features is in focus at any given moment. A second option in line with the discipline of physics is the name "Fielding." If God can be understood as linked to a field, which in turn generates particles with which we are familiar, maybe Fielding is an appropriate name for God. The advantages are these: as a surname it is not gender-specific and it implies that God boasts some type of personality. As a noun it refers to a field and as a verb it entails that God is actively in the business of working in and/or through fields. If this will be found to be consistent with the physics of fields, then it would constitute a scientific improvement over terms like "World Soul" and "Universal Mind."

My final submission comes from an affection for acronyms. This has the advantage of combining a number of different aspects of divinity in a short-hand way; the disadvantage is that it can be cumbersome. Only the reader can decide. The name is P.R.I.T.C.H.A.R.D., which stands for Prime Reality (with) Immanent (and) Transcendent Components Hopefully Attaining Real Differences. "Hopefully" because, in the Whiteheadian strategy, there is no guarantee that God's ideals will be grasped; and "attaining" (or perhaps even "assembling") since God extends efforts through God's own final cause to actualize God's aims. The only stipulation, of course, is that in the process scheme God cannot do so unilaterally. And "differences" need to be "real" because they must matter to us and have some macro-level impact. Once again, as a surname it is non-gender-specific and points to a personality.

This offering might not obtain wide acceptance, but at least it presents alternative designations. It is one attempt to liberate ourselves from the burden of employing an appellation that, in the eyes of some, no longer serves a useful purpose. And with this I have completed one ecumenical task.

APPENDIX TWO

Time for a Change

THE CAVEAT APPLICABLE TO THE FIRST APPENDIX, namely its nonessential character, is doubly the case for this second one. This is because, breaking with tradition, I have elected to mix fiction with non-fiction. It is not customary to include a piece, in this instance a short story, on speculative fiction in an otherwise non-fictional account that attempts to present and defend a metaphysical proposal. So much for custom. The submission of this tale, which incidentally makes no mention of divinity, is offered purely for the reader's enjoyment. Consider it a gift of appreciation for those who have stuck with my treatise throughout.

Back in the days when teleportation was all the rage, it seemed to be the upper class thing to do on weekends to jettison the particles of one's body to a distant place. Not so much because that's where you really wanted to be, but just for the sheer exhilaration of conquering what was formerly considered a physical barrier. "Instantaneous location displacement," or I.L.D. for short, is the fancy name that was given it.

"A Concorde Jet is the fastest way to travel," they used to say a long time ago (within the confines of the earth's atmosphere, of course). Jumbo jets and other aircraft now fill junkyards and scrapheaps. The lower classes still use them as an inexpensive way to get around, but the use of planes has diminished, and they definitely are not the mode of transportation of choice. It's a shame nobody thought of making them fully recyclable during their heyday. What were they thinking? That planes were the end of the line technologically, and that there were no more engineering marvels to ensue?

Hardly! Talk about a revolution! Teleportation devices not only became a time-saver, but the anticipated bugs of faulty particle realignment were never an issue. One of the happy quirks of science, I guess. Fortunately, no one had to suffer the indignity of physical deformation after an I.L.D. "trip." The particles just seemed to know their proper order. It's a matter for the physicists, and the biologists too I suppose, to ponder. As for me, I'm too busy basking in the glorious glow of the fact that it works. I imagine that makes me an unabashed pragmatist. So be it.

But the revolution did not arrive free of logistical problems. One could not simply enjoy the fruits of the new technology in a free-wheeling way. Just think of the mayhem that would result if two or more people projected themselves onto the same coordinates. The physicists are correct when they say that no two objects can occupy the same space at the same time. This has been known for centuries and, inconvenience aside, it is no less true even today.

This is why the government chose to step in and regulate the comings and goings of travellers. This was not an attempt to limit the freedom of its citizenry, nor was it perceived as such. It was merely the best way to keep hospital wards from potentially filling up. Think of it as the equivalent of the twentieth-century task of air-traffic control at airports. This was a necessary service about which nobody objected. Similar thing here. Already existing airports were also selected as sites for the set up of these new teleportation devices. Might as well save on resources and use what you've already got, was their reckoning. Schedules for "flights" were then coordinated with other "teleports," hopefully to eliminate the possibility of casualties. It would have been a major embarrassment to have "mid-air" or "on-land collisions" when they are completely unnecessary and when there are no real time constraints. You can't get much faster than instantaneous!

The business community stood to benefit the most from this new mode of transportation. Conference calls became a thing of the past; you just arrange a meeting and—Europa!—all the participants could make it. No longer was traffic a legitimate excuse for tardiness or absence. (No one claimed that teleportation was entirely free of negative aspects.)

Yet these devices are far too expensive for ordinary individuals to own—much to the contentment of governments. This is like the situation, or so I am told, for the wealthy class of the twentieth century. It was well within the possibility for those with the means and the extravagance to own and operate a private jet. Not here. Large corporations, though, could have been in a position to have a stake in them, keeping alive for governments the age old issue of whether to privatize.

This didn't appear to bother anybody, however. Nobody, that is, except my friend Edgar. Edgar was in engineering research and development. Now retired, he tinkers about on his own, working on what he says "the fools" would not trouble themselves with. Now you might think that he was simply trying to build a teleportation device for himself, on the sly. No, Edgar was not satisfied with teleportation. He wanted to time travel.

"Same principle," he would say. "After all, time is space according to relativity theory. There is a space-time continuum, not space and time. The two can be distinguished but never completely separated. In teleportation you alter the space in between the points of departure and arrival. Time is also affected thereby. Bend one and you bend the other."

This is as far as I got with his analysis. Anything beyond that was using hieroglyphics for symbols as far as I was concerned. Anyway, at the end of twelve years working on his invention, he thinks he has got it just right. As right as it can be, that is, without it ever having been subject to a trial run.

"Kyle," he would tell me, his big brown eyes widening as he spoke, "we're going to make history. And I do mean make it, because by entering any age from a different period than the one which we now occupy, it is likely that we will change the outcome of events."

Now you will notice that he used the term "we" here. The reason for this is, well, who do you think is not going to take this invention on its maiden voyage, for its initial spin? Somebody has to stay behind in the laboratory to work the controls on the transmitting console so that the "driver" can receive the signals. That would be me. Neither of us considered me to be qualified for the former job. If anyone was going to be the first time traveller, it was Ed.

We didn't have far to look for a "destination" either. It would be the past of this very town. He hadn't yet got to the advanced stage where travel into the future would be likely, nor could he as yet combine teleportation with what he claimed his machine could do, namely "temporate." He could not relocate or teleport to a different place as well as time. What is more, the time in which he could travel is restricted to the past.

"Those other abilities may still be a long way off," he conceded.

As for any bugs in the system, Ed referred to what was known as chaos theory.

"You have to be careful how you interact in a different time period," he warned. "Someone from another time is an intruder and can potentially upset a delicate balance of factors that have produced the space-time we now inhabit. The slightest bit of interference from such a foreign element can have a ripple effect that will be sent cascading down through time and affect us in the here-now."

"Then why take the chance?" I ventured.

"Because there are some things that are worth changing," he replied.

"Assuming you align all the parts just right, I suppose. But whose benefit will this be for, yours or others?" It sounded like I was interrogating him.

"Hopefully both," he responded, and left it at that.

Ed climbed into the machine, sat down and buckled himself in.

"You expecting a bumpy ride?" I inquired, attempting to lessen the tension.

"If I did, I'd wear a helmet," he retorted.

He had given me instructions as to what to push and turn and when to do so. Everything else was preset to the correct position.

"Off I go then," was the last I heard him say. He faded from view, as one would in a teleportation device. I was anticipating something a little more dramatic, I guess. Perhaps the temporal equivalent of a sonic boom. But then that doesn't happen when teleporting, so why should it during "temporating"? It hurt to reflect upon it any further, so I decided to stop.

What I did experience, however, was a rumbling underneath my feet for what must have been several seconds. Immediately after it subsided, Ed returned.

"Well that was quick!" I exclaimed, celebrating the victorious temporal conquest.

"Maybe for you," he said. "I hope my efforts paid off."

Shortly thereafter I noticed something different about Ed, in his appearance I mean. I couldn't put my finger on it at first, but then it struck me that he was sporting a full beard.

"Evidently you must have been gone long enough from this time frame to grow a beard," I mused.

"What do you mean?" he asked. "I've always had one."

"I distinctly recall you as always having been clean shaven," I insisted.

"You must be mistaken," he said.

"Do you have a photo of yourself lying around somewhere so that we could verify this?" I asked.

"No, oddly enough I have never owned a camera," he admitted. "But if it makes you feel any better, for future reference run home and get yours before I take my next trip."

"Which is when?" I inquired.

"As soon as you snap the shot," he responded. "Anything happen while I was gone? And how much time elapsed?"

"Less than a minute," I estimated. "Of my time here, that is. Oh, and there was a rumbling, like a low-grade earthquake, for a short while. What was that all about?"

"I expected as much." Ed did not seem surprised. "Let me know if you notice anything different outside. Now go and get your camera."

I went and did as I was told. Nothing appeared different, at least at first glance. Then I detected a difference in the skyline of our town. There used to be a building there. More accurately, it was still there, only shorter, maybe half its former size. I looked around for dust and debris, anticipating that the rumble I felt resulted in the toppling of half of this structure. I didn't find any.

When I got back I reported on what I saw.

"That's what I was hoping." He sounded relieved. "Not only can this machine take me back and forth, but with a little bit of adjustment in the events of the past, one can bring about desired effects for the present."

"You orchestrated a shorter building?" I ventured to ask.

"Yes, it's amazing what a little incentive can do," he sneered.

"That was the entire purpose of your trip?"

"Relax. You start off with the small and then advance to the large."

"The building was a small thing to you?" I questioned his motives.

"To me it was a big thing—too big. It blocked out the sun for me and cast a shadow on my workshop much too early in the day. So I resolved to make a big thing a shorter thing. A small matter, really."

"How did you pull it off?"

"I simply went back a decade in time and had a talk with the city official in charge of it. I mentioned to him that I knew of some illegitimate business practices he was engaged in and informed him of my option to bring them to light if the building permit did not reflect the change in construction plans I requested."

"You threatened him!"

"Merely pointing out my interests. Effective, though, wouldn't you say?"

"Perhaps so, but is that all you made your invention for—the fine-tuning of your surroundings to suit your own tastes?"

"No, that was just a warm-up exercise. I have something more in mind. My next trip will be to the time prior to my parents' wedding. My father gave my mother no end of anxiety with all his drinking and abuse. My uncle, on the other hand, would have been a much better match for my mother. We would have had a happier home life without my father under the same roof. I'm going to see what I can do about that."

"You are going to play historical matchmaker? What about the interference you were talking about? Tamper with the initial conditions of a system, you said, and predictability is lost after a few iterations. Remember?"

"According to my calculations, this is a risk worth taking. Now go ahead and take my picture."

Reluctantly, I aimed the camera and took a head and shoulders shot of Ed. The kind you would need for a passport. I speculated that, once time travel becomes standard fare, maybe there would be some kind of office around to regulate it and for which you would need some type of authorization—all the clearance that any historical tinkering would require. Perhaps that would prevent the misuse of this new capability. As it stood, I had misgivings about the whole affair and showed little enthusiasm.

The photograph developed right before my eyes and portrayed an unflattering image of a worn but resolute engineer.

"It's time to correct my past," was all he said.

It was futile to try and stop him or even reason with him at this point. There was a determination in his eyes. I only hoped that he had taken the time to check his calculations for errors.

I held the snapshot as I watched him fade, fully expecting him to return any moment. I dutifully twisted and turned, pushed and pulled the buttons and levers at the approved times. I hoped the changes he was about to make—what am I saying?—will have made by now, were for the better. I wasn't convinced that would be the case.

There were no rumblings this time, only a slight tingling sensation in the ends of my fingers. The sky did not turn dark, nor did anything blink out of existence. I would still be there to greet the intrepid traveller upon his arrival. A good thing too, else who would have worked the controls?

The machine materialized once again, or "retemporalized" I suspect. It's all so confusing.

Everything seemed normal, perfectly ordinary. At least to begin with. The traveller who then got unbuckled from the seat and stepped out of the machine reminded me that her name was Amanda.

She had blue eyes.

Just like in her picture.

Notes

1. Owens, as she admits, has drawn these words from Owen Barfield.
2. Hooykaas 1969, 202 n.1, speaks here of Adam Sedgwick.
3. Lovelock 1988, 42. All emphases in this study are on the part of the author(s) cited unless otherwise specified.
4. This thought is taken from Kuhn's postscript (p. 208) written seven years after the publication of the first edition.
5. This is the publisher's caption appearing on the front cover of *The study of man*.
6. These three quotations are taken from Nicholas Wolterstorff, *Reason within the bounds of religion* (Grand Rapids, MI: Eerdmans, 1976), pp. 16, 49 and 98, respectively.
7. Third edition (Cambridge, MA: Harvard University Press, 1957).
8. For a more in-depth account of the Whiteheadian process framework the reader is directed to my previous volume *How in the world does God act?* (University Press of America, 2000).
9. Whitehead cautions that "in the main the philosophy of organism is a recurrence to pre-Kantian modes of thought" (Griffin & Sherburne, eds., 1978, xi); and also warns that "Religion will not regain its old power until it can face change in the same spirit as does science. Its principles may be eternal, but the expression of those principles requires continual development" (1967b, 189).
10. It would have been advantageous had Barbour provided these references whereby one could follow up on the evidence in support of his view. See also Stephen Jay Gould, *Hen's teeth and horse's toes* (New York: Norton, 1983), p. 171.
11. Consult my two articles in *The Journal of Religion and Psychical Research*: The compatibility of process thought with the psychical di-

mension. 23(2): 67-72, April 2000; and God and the new metaphysics. 24(2): 64-9, April 2001.
12. David Ray Griffin, *Parapsychology, philosophy, and spirituality: A postmodern exploration* (Albany, NY: SUNY Press, 1997).
13. Rupert Sheldrake, *Dogs that know when their owners are coming home: And other unexplained powers of animals* (New York: Three Rivers Press, 1999), and *The sense of being stared at: And other aspects of the extended mind* (New York: Crown, 2003).
14. Both quotations are taken from Lewis Carroll, *The annotated Alice* (New York: New American Library, 1974), pp. 269-70.
15. Ibid., pp. 191, 196-7. Both are contained in the poem *Jabberwocky*.

Bibliography

Ayala, Francisco J. 1998. Darwin's devolution: Design without designer. In Robert John Russell, William R. Stoeger, S.J., and Francisco J. Ayala, eds. *Evolutionary and molecular biology: Scientific perspectives on divine action*. Vatican City State: Vatican Observatory & Berkeley, CA: Center for Theology and the Natural Sciences: 101-16.

Barbour, Ian G. 1966. *Issues in science and religion*. New York: Harper & Row.

--------. 1990. *Religion in an age of science*. San Francisco: HarperCollins.

--------. 1997. *Religion and science: Historical and contemporary issues*. San Francisco: HarperCollins.

--------. 1998. Five models of God and evolution. In Robert John Russell, William R. Stoeger, S.J., and Francisco J. Ayala, eds. *Evolutionary and molecular biology: Scientific perspectives on divine action*. Vatican City State: Vatican Observatory & Berkeley, CA: Center for Theology and the Natural Sciences: 419-42.

Barrow, John D. 1995. *The artful universe: The cosmic source of human creativity*. Toronto: Little, Brown & Co.

Barrow, John D., and Frank J. Tipler. 1988. *The anthropic cosmological principle*. Oxford: Oxford University Press.

Becker, Robert O., and Gary Selden. 1985. *The body electric: Electromagnetism and the foundations of life*. New York: Quill/William Morrow.

Bernstein, Jeremy. 1973. *Einstein*. Glasgow: Fontana.

Birch, Charles. 1998. Neo-Darwinism, self-organization, and divine action in evolution. In Robert John Russell, William R. Stoeger, S.J., and Francisco J. Ayala, eds. *Evolutionary and molecular biology: Scientific perspectives on divine action*. Vatican City State: Vatican

Observatory & Berkeley, CA: Center for Theology and the Natural Sciences: 225-48.
Bohm, David. 1980. *Wholeness and the implicate order*. London: Routledge & Kegan Paul.
--------. 1986. Time, the implicate order and prespace. In David Ray Griffin, ed. *Physics and the ultimate significance of time: Bohm, Prigogine and process philosophy*. Albany, NY: SUNY Press: 177-208.
--------. 1988. Postmodern science and a postmodern world. In David Ray Griffin, ed. *The reenchantment of science: Postmodern proposals*. Albany, NY: SUNY Press: 57-68.
Bohm, David, and Basil J. Hiley. 1993. *The undivided universe: An ontological interpretation of quantum theory*. New York: Routledge.
Bohm, David, and F. David Peat. 1987. *Science, order, and creativity*. Toronto: Bantam.
Brecht, Bertolt. 1960. *The life of Galileo*. Translated by Desmond I. Vesey. London: Methuen & Co.
Brummer, Vincent, ed. 1991. *Interpreting the universe as creation: A dialogue of science and religion*. Kampen, The Netherlands: Kok Pharos.
Capra, Fritjof. 1983. *The Tao of physics*. Rev. edn. London: Fontana.
Clayton, Philip D. 1997. *God and contemporary science*. Grand Rapids, MI: Eerdmans.
Clifford, Anne M. Darwin's revolution in the *Origin of species:* A hermeneutical study of the movement from natural theology to natural selection. In Robert John Russell, William R. Stoeger, S.J., and Francisco J. Ayala, eds. *Evolutionary and molecular biology: Scientific perspectives on divine action*. Vatican City State: Vatican Observatory & Berkeley, CA: Center for Theology and the Natural Sciences: 281-302.
Davies, Paul. 1983. *God and the new physics*. London: J.M. Dent &Sons.
--------. 1992. *The mind of God: The scientific basis for a rational world*. Toronto: Simon & Schuster.
--------. 1998. Teleology without teleology: Purpose through emergent complexity. In Robert John Russell, William R. Stoeger, S.J., and Francisco J. Ayala, eds. *Evolutionary and molecular biology: Scientific perspectives on divine action*. Vatican City State: Vatican Observatory & Berkeley, CA: Center for Theology and the Natural Sciences: 151-62.

--------. 2000. *The fifth miracle: The search for the origin and meaning of life.* New York: Simon & Schuster.
Devereux, Paul, John Steele, and David Kubrin. 1989. *Earthmind: A modern adventure in ancient wisdom.* Cambridge, MA: Harper & Row.
Drees, Willem B. 1990. *Beyond the big bang: Quantum cosmologies and God.* Lasalle, IL: Open Court.
Dugatkin, Lee Alan. 2000. *The imitation factor: Evolution beyond the gene.* New York: Free Press.
Evans, C. Stephen. 1982. *Philosophy of religion: Thinking about faith.* Downers Grove, IL: InterVarsity Press.
Ferre, Frederick. 1988. Religious world modeling and postmodern science. In David Ray Griffin, ed. *The reenchantment of science: Postmodern proposals.* Albany, NY: SUNY Press.
--------. 1993. *Hellfire and lightning rods: Liberating science, technology, and religion.* Maryknoll, New York: Orbis.
Flew, Antony. 1984. *Darwinian evolution.* London: Paladin.
Gerhart, Mary, and Alan Russell. 1984. *Metaphoric process: The creation of scientific and religious understanding.* Fort Worth: Texas University Press.
Gilkey, Langdon. 1965. *Maker of heaven and earth: A study of the Christian doctrine of creation.* New York: Anchor.
--------. 1979. *Message and existence: An introduction to Christian theology.* New York: Seabury Press.
--------. 1981. *Religion and the scientific future: Reflections on myth, science and theology.* Macon, GA: Mercer University Press.
--------. 1985. *Creationism on trial: Evolution and God at Little Rock.* Minneapolis, MN: Winston Press.
--------. 1989. Nature, reality and the sacred: A meditation in science and religion. *Zygon* 24: 283-98, September.
Globus, Gordon G. 1987. Three holonomic approaches to the brain. In Basil J. Hiley and F. David Peat, eds. *Quantum implications: Essays in honour of David Bohm.* New York: Routledge & Kegan Paul: 372-85.
Gould, Stephen Jay. 1980. *The panda's thumb: More reflections in natural history.* New York: Norton.
--------. 1983. *Hen's teeth and horse's toes: Further reflections in natural history.* New York: Norton.
--------. 1987a. *Time's arrow, time's cycle: Myth and metaphor in the discovery of geological time.* Cambridge, MA: Harvard University Press.

--------. 1987b. *An urchin in the storm: Essays about books and ideas*. New York: Norton.

--------. 1989. *Wonderful life: The Burgess Shale and the nature of history*. New York: Norton.

--------. 1993. *Eight little piggies: Reflections in natural history*. London: Jonathan Cape.

--------. 1995. *Dinosaur in a haystack: Reflections in natural history*. New York: Crown.

--------. 1996. *Full house: The spread of excellence from Plato to Darwin*. New York: Three Rivers Press.

--------. 1999. *Rocks of ages: Science and religion in the fullness of life*. New York: Ballantine.

Griffin, David Ray. 1988a. Introduction: The reenchantment of science. In David Ray Griffin, ed. *The reenchantment of science: Postmodern proposals*. Albany, NY: SUNY Press.

--------. 1988b. Of minds and molecules: Postmodern medicine in a psychosomatic universe. In David Ray Griffin, ed. *The reenchantment of science: Postmodern proposals*. Albany, NY: SUNY Press.

--------. ed. 1989a. *Archetypal process: Self and divine in Whitehead, Jung, and Hillman*. Evanston, IL: Northwestern University Press.

--------. 1989b. *God and religion in the postmodern world: Essays in postmodern theology*. Albany, NY: SUNY Press.

Griffin, David Ray, and Donald W. Sherburne. eds. 1978. *Process and reality: Corrected edition*. New York: Free Press.

Harman, Willis W. 1988. The postmodern heresy: Consciousness as causal. In David Ray Griffin, ed. *The reenchantment of science: Postmodern proposals*. Albany, NY: SUNY Press.

Hasker, William. 1983. *Metaphysics: Constructing a world view*. Downers Grove, IL: InterVarsity Press.

Haught, John F. 1998. Darwin's gift to theology. In Robert John Russell, William R. Stoeger, S.J., and Francisco J. Ayala, eds. *Evolutionary and molecular biology: Scientific perspectives on divine action*. Vatican City State: Vatican Observatory & Berkeley, CA: Center for Theology and the Natural Sciences: 393-418.

--------. 2000. *God after Darwin: A theology of evolution*. Boulder, CO: Westview Press.

Hawking, Stephen W. 1988. *A brief history of time: From the big bang to black holes*. Toronto: Bantam.

Henry, Granville C., and Robert J. Valenza. 1994. The principle of affinity in Whiteheadian metaphysics. *Process Studies* 23(1): 30-49, Spring.

Hooykaas, Reijer. 1969. *Law and divine miracle: A historical-critical study of the principle of uniformity in geology, biology, and theology.* Leiden, The Netherlands: E.J. Brill.
Johnson, Philip E. 1991. *Darwin on trial.* Downers Grove, IL: InterVarsity Press.
Kirk, James Ernest. 1991. Organicism as reenchantment: Whitehead, Prigogine, and Barth. Ph.D. dissertation, University of Texas at Dallas.
Kuhn, Thomas S. 1970. *The structure of scientific revolutions.* Chicago: University of Chicago Press.
Leslie, John. 1989. *Universes.* New York: Routledge.
Lovelock, James E. 1979. *Gaia: A new look at life on earth.* Oxford: Oxford University Press.
--------. 1988. *The ages of Gaia: A biography of our living earth.* New York: W.W. Norton & Co.
Lukas, Mary, and Ellen Lukas. 1977. *Teilhard: The man, the priest, the scientist.* Garden City, NY: Doubleday & Co.
MacKay, Donald M. 1974. *The clockwork image.* Downers Grove, IL: InterVarsity Press.
Morris, Simon Conway. 1998. *The crucible of creation: The Burgess Shale and the rise of animals.* New York: Oxford University Press.
Owens, Virginia Stem. 1983. *And the trees clap their hands: Faith, perception, and the new physics.* Grand Rapids, MI: Eerdmans.
Peacocke, Arthur. 1993. *Theology for a scientific age: Being and becoming—natural, divine and human.* Minneapolis, MN: Fortress Press.
--------. 1998. Biological evolution—A positive theological appraisal. In Robert John Russell, William R. Stoeger, S.J., and Francisco J. Ayala, eds. *Evolutionary and molecular biology: Scientific perspectives on divine action.* Vatican City State: Vatican Observatory & Berkeley, CA: Center for Theology and the Natural Sciences: 357-76.
Peat, F. David. 1997. *Infinite potential: The life and times of David Bohm.* Reading, MA: Addison-Wesley.
Playfair, Guy Lyon, and Scott Hill. 1979. *The cycles of heaven: Cosmic forces and what they are doing to you.* London: Pan.
Polanyi, Michael. 1946. *Science, faith and society.* Toronto: University of Toronto Press.
--------. 1959. *The study of man.* Chicago: University of Chicago Press.
--------. 1967. *The tacit dimension.* Garden City, NY: Anchor/Doubleday.

Polkinghorne, John. 1989. *Science and providence: God's interaction with the world*. Boston: Shambhala.

Randall, John L. 1977. *Parapsychology and the nature of life*. London: Abacus.

Ratzsch, Del. 1986. *Philosophy of science: The natural sciences in Christian perspective*. Downer's Grove, IL: Inter-Varsity Press.

Russell, Robert John. 1993. Introduction. In Robert John Russell, Nancey Murphy, and C.J. Isham, eds. *Quantum cosmology and the laws of nature: Scientific perspectives on divine action*. Vatican City State: Vatican Observatory & Berkeley, CA: Center for Theology and the Natural Sciences: 1-32.

--------. 1998. Special providence and genetic mutation: A new defense of theistic evolution. In Robert John Russell, William R. Stoeger, S.J., and Francisco J. Ayala, eds. *Evolutionary and molecular biology: Scientific perspectives on divine action*. Vatican City State: Vatican Observatory & Berkeley, CA: Center for Theology and the Natural Sciences: 191-223.

Sheldrake, Rupert. 1987. *A new science of life: The hypothesis of formative causation*. 2nd ed. London: Paladin.

--------. 1988a. The laws of nature as habits: A postmodern basis for science. In David Ray Griffin, ed. *The reenchantment of science: Postmodern proposals*. Albany, NY: SUNY Press.

--------. 1988b. *The presence of the past: Morphic resonance and the habits of nature*. London: Collins.

--------. 1994. *The rebirth of nature: The greening of science and God*. Rochester, VT: Park Street Press.

Sherburne, Donald W. ed. 1981. *A key to Whitehead's* Process and reality. Chicago: University of Chicago Press.

Stoeger, William R., S.J., 1998. The immanent directionality of the evolutionary process, and its relationship to teleology. In Robert John Russell, William R. Stoeger, S.J., and Francisco J. Ayala, eds. *Evolutionary and molecular biology: Scientific perspectives on divine action*. Vatican City State: Vatican Observatory & Berkeley, CA: Center for Theology and the Natural Sciences: 163-90.

Talbot, Michael. 1992. *The holographic universe*. New York: Harper-Perennial.

Teilhard de Chardin, Pierre. 1965. *The phenomenon of man*. London: Fontana.

Templeton, John, and Robert Hermann. 1988. *The God who would be known: Divine revelations in contemporary science*. San Francisco: Harper & Row.

Torrance, Thomas F. 1984. *Transformation and convergence in the frame of knowledge: Explorations in the interrelatedness of scientific and theological enterprise*. Grand Rapids, MI: Eerdmans.

van Huyssteen, Wentzel. 1989. *Theology and the justification of faith: Constructing theories in systematic theology*. Translated by H.F. Snijders. Grand Rapids, MI: Eerdmans.

Watson, Lyall. 1980. *Lifetide*. London: Hodder & Stoughton.

--------. 1992. *The dreams of dragons: An exploration and celebration of the mysteries of nature*. Rochester, VT: Destiny.

--------. 1995. *Dark nature*. London: Hodder & Stoughton.

Weinberg, Steven. 1992. *Dreams of a final theory*. New York: Pantheon.

Whitehead, Alfred North. 1967a. *Adventures of ideas*. New York: Free Press.

--------. 1967b. *Science and the modern world*. New York: Free Press.

--------. 1974. *Religion in the making*. New York: Meridian/New American Library.

Wolterstorff, Nicholas. 1976. *Reason within the bounds of religion*. Grand Rapids, MI: Eerdmans.

Wright, Richard T. 1989. *Biology through the eyes of faith*. San Francisco: Harper & Row.

Zubiri, Xavier. 1981. *Nature, history, God*. Lanham, MD: University Press of America.

Zukav, Gary. 1980. *The dancing Wu Li masters: An overview of the new physics*. Toronto: Bantam.

Index

actualistic gradualism, 30
Adams, D., 5-6
adaptive variations, 73, 75, 76, 78, 89, 91-92, 94
   Gould's view, 98
   see also natural selection
aerobes and anaerobes, 106, 109
after-the-fact argument, 84
Agar, W.F., 123
analogies
   football, 65-72
   sports statistics, 100-102, 128-29
   watching TV with pet, 50-51
anthropic principle, 42-45, 82, 83, 85, 165
   Lovelock's view, 114
   nontheistic versions, 90
   strong form, 44, 45, 83
   weak form, 44-45
Aquinas, T., 15, 80, 90
arrow of time, *see* directionality
Aspect, A., 152
atmosphere as control system, 106-107
Ayala, F.J., 75, 78

Barbour, I.G., 61, 77, 79, 80, 81, 151, 160-61
Barrow, J., 29, 37, 38, 42, 44, 46, 90
Barrow, J.D., 112
Barth, K., 174
baseball statistics and evolution, 100-102
Bateson, G., 56

Becker, R.O., 130-32, 137, 168, 169
Bell, J.S., 41, 152
Bergson, H., 125
Berkeley, G., 16, 160
Berman, M., 53-56
Bernstein, J., 34
Bible, 172-74
Big Bang, 6, 33, 35, 46-47, 74
Birch, C., 76, 148
bird migrations, 127
Bohm, D., 41, 53, 149-55, 156, 159, 160
Bohr, N., vs. Einstein, 36-42, 51
Brecht, B., 8
Brummer, V., 26
Burgess Shale, 102
butterfly migrations, 167

canalization of development, 119-20, 126
"Cat Paradox," 39-40
catastrophism, 30
"causal efficacy," (Whitehead), 140, 141
"causal interpretation" (Bohm), 153
causality of past, 138-45
causality principle, 30
cause and effect in Einstein's world, 34-35
chance and natural selection, 78-79
chreodes, 119, 120, 122
circadian patterns, 131, 168
classical view of God, 80, 88, 89, 93
Clayton, P.D., 161
Clifford, A.M., 77

Cobb, J.B., Jr., 63
collective unconscious, 136
complementarity principle, 41
complexity, increasing, 75-76, 86, 100
Comte, A., 18
consciousness, 43-44, 82, 133, 158
constants of nature, 42-43, 83, 90
contingency of universe, 47-48
convergence, 104, 133-34
Conway Morris, S., 91, 102-104
Copernicus, N., 20-21
creativity for Whitehead, 62-63, 66, 70, 89
crystals
   self-replicating powers, 83, 128
   synthesis of new crystals, 121-22
cultural evolution, 129
cyclic patterns of people and planet, 170

Darwin, C., 2, 57, 129, 167-68
   natural selection, 72, 73-75, 77-78
   on purpose and progress, 77-78, 86
Darwin, E., 73
Davies, P., 5, 113
   design in universe, 42, 43, 44, 47, 48
   quantum systems, 33-34, 35-36, 37-38, 40-41
   science and religion, 49, 50, 64, 165
Descartes, R., 14-15, 54
design, 82-89, 90
   God as Designer, 87-88, 90
   see also anthropic principle; teleology
destructive elements in the world, 87
determinism, 34-35, 38-39, 40, 41, 48, 51
Devereux, P., 169
dinosaurs, 91
dipolarity, 65
directionality, 35-36, 52, 74, 75, 82
   and evolution, 75-78, 82, 94-95, 97-104, 114, 134-37
   see also teleology
divinity, *see* God
DNA, 115-16, 126-28
dodo, extinction by humans, 92
Draper, J.W., 10, 11, 53

Drees, W., 45-46, 48, 49, 52, 64, 65
dualism, 6-7
   Bohm's explicate and implicate orders, 155
   Cartesian, 14-15, 58-59, 160
   in vitalist theories, 117
   Whitehead's universe, 58-59, 61, 65
Dugatkin, L.A., 129

Einstein, A., 18, 25, 34, 35, 151-52, 153
   determinism of, 34-35, 38-39, 40
   vs. Bohr, 36-42, 51
Eldridge, N., 76
electromagnetic radiation (EMR), 130-31, 171
emergent properties, 63, 64
emergentism, 149
empiricism, 15, 16, 38, 140
   "deep empiricism", *see* organicism opposed to metaphysics, 17-18
empowerment, 148
entelechy, 117
entities (in process thought), 63-64, 71, 80, 81, 88, 89, 94
EPR thought-experiment, 38-39, 41, 151-52
Evans, C.S., 25-26
events (Whitehead), 66, 67, 69
evil, 147-48
evolutionary theory, 72, 73-95
   baseball statistics as analog, 100-102
   and cosmic design, 82-89
   and directionality, 75-78, 82, 94-95, 97-104
   increase in complexity, 75-76, 86, 99-100
   indeterminacy, 75, 78-80
   opposed to special creation, 74
   pain and waste, 80, 87, 89, 93
   process view, 80-82, 88-89, 94-95
   and progress, 76, 85-86, 88-89, 95, 99-104
   and purpose, 75, 77, 79-80, 90-95
   randomness, 75, 78, 98, 99
explicate and implicate orders, 150-51, 152, 154-55
extinction of species, 91-92, 93, 94

falsifiability, 19, 20
Ferre, F., 32, 62
field of vision, limited, 50-51
fields
  Bohm's concept, 154, 156-60
  morphogenetic, 118-19, 120, 126, 128, 132, 156-60
football analogy for process thought, 65-72
formative causation (Sheldrake), 119-30
four-minute mile, 128-29
framing of history, 9-10
future, 135-36

Gaia hypothesis, 105-114, 137
  as teleological, 110-11
  view of pollution, 108-110
Galileo, 8, 11
general relativity, 34, 35, 45
genetics, 115-16, 126-28, 166, 167
geology, approaches to, 29-31
Gerhart, M., 49-50
Gilkey, L., 26-29, 48, 49
Globus, G.G., 154
God
  and the Bible, 172-74
  changing pictures of, 2-3
  classical view, 80, 88, 89, 93
  as designer, 87-88, 90
  and evil, 147-48
  ideal, 64, 71, 81, 88-89, 135-36, 140
  involvement in world, 79-89, 93-95, 148, 156-62
  kenotic model, 88, 148
  name, 176-79
  and scientific knowledge, 15
  traces in the cosmos, 43-45, 51-52, 82-83, 90
  Whitehead's view *see* Whitehead, A.N.
  see also progress; purpose; teleology
Gould, S.J., 10, 11, 76, 91, 92
  ideas on purpose, 84, 86, 87
  opposition to directionality, 85, 97-104
Griffin, D.R., 57, 58, 59-62, 144, 145, 146-47, 174

habits (Sheldrake's view), 122, 124
Harman, W.W., 62
Hartshorne, C., 88
Hasker, W., 149
Haught, J.F., 80
Hawking, S., 26
Heisenberg, W., 37, 55
hemoglobin folding, 118, 119
Henry, G.C., 124
Heraclitus, 145
heredity, *see* DNA; inheritance (Sheldrake's view)
Hermann, R., 150
Hiley, B.J., 152, 154, 155
Hill, S., 168
Hitchhiker's Guide to the Galaxy (Adams), 5-6
homeostasis, 108
Homo sapiens, 91
Hooykaas, R., 30-31, 53
Hoyle, F., 165
humans
  evolution of, 91
  as purpose of universe, *see* anthropic principle
Hume, G., 16
Hutton, J., 87
hypothesis and reality, 8-9, 19-21

ideal, God's aim for universe, 64, 71, 81, 88-89, 94-95
idealism, 6, 12, 16, 160
ideas
  maps, frames for reality, 6, 96, 104, 156, 175
  observer's ideas shape reality, 8-9, 19-21
  paradigms, 13, 19-21, 24
ideation, 167
imagination in theory-formation, 24, 25, 29
immanence, 54, 59
implicate and explicate orders, 152-53, 154-55, 159
Incarnation, Christian doctrine, 156-57
indeterminacy
  of natural world, 75, 78, 79-80

indeterminacy *(continued)*
   of quantum world, 33-34, 35, 37-38, 40, 41-42, 51, 78-79, 80, 161, 164, 165
inheritance (Sheldrake's view), 120-21, 122-23
instincts (Sheldrake's view), 116, 122
instrumentalism, 18, 49
intelligibility of universe, 47, 48
interventionism, 80, 94, 157.
   See also God
intuition, 23, 29, 136

Johnson, P.E., 20, 21
Jung, C.G., 136

Kant, I., 16-17, 54-55
kenotic model of God, 88, 148
Kirk, J.E., 62, 63, 64
knowledge as personal, 22-23
Kohlberg, L., 173
Kubrin, D., 169
Kuhn, T., 2-3, 19-22, 23, 24

Lamarck, J.P. de, 120, 128
Langer, S., 55
Laplace, P., 15
laws of nature, 36, 123-24
learning, inheritance of, 122-23
Leslie, J., 42-43, 46
"limit questions," 29
limits
   to human perception, 50-51
   to observation, 33, 37
Locke, J., 15
Lovelock, J., 8, 47, 105-114

Malthus, T.R., 73
many-worlds theory, 165
maps, frames for reality, 6, 96, 104, 156, 175. *See also* paradigms
materialism, 6, 12, 58, 60, 160
McDougall, W., 122-23
McFague, S., 13
mechanistic view, 15, 34-35, 54, 174
   of living forms, 116-17, 125, 126
memory, 66, 139-40, 171
Mendel, G., 75

metaphoric process, 49-50
metaphors, 48
metaphysics opposed by empiricism, 17-18
methodology or metaphysics, 42
mind-brain model, 51-52, 161-62
misplaced concreteness, fallacy, 9, 58
monism, 6, 160
morphic resonance, 119-20, 121, 128
morphogenetic fields, 118-19, 120, 126, 128, 132
Morris, S.C., *see* Conway Morris, S.
Mother Earth, 169
mutations, 78, 97, 98
mystery religions, 12
myths, 27, 28, 48, 49

"natural gaps," 79
natural selection, 72, 73, 75, 76-77
   Gould's view, 99-100
   in process thought, 89, 93
   and teleology, 27-28, 77-78, 92-93
natural theology, 90
Newton, I., 15, 54, 73
nontheistic arguments for purpose, 90-93
noosphere, 135, 136

observation
   God's existence inferred from, 42-45, 82-85, 90
   limits to, 33, 37
   paradigms, 13, 19-21
   and reality, 16, 17-18
   role of the observer, 22-23, 24, 28, 31
   and theory, 14, 19, 24-25, 129-30
observer interference, 41-42, 55
Omega point, 135-36
openness of quantum and evolutionary worlds, 78-79, 136
organ donors and recipients, 171
organic realism (Whitehead), 57-62
organicism, 60-62, 116, 117-18, 126
organizing center of society, 65, 67-68
Owens, V.S., 8, 9, 22, 34, 35, 42

Paley, W., 90
panentheism vs pantheism, 65, 146

Index

panexperientialism, 6, 61, 145
Pannenberg, W., 24
pantheism, 65, 160
paradigms, 3, 13, 19-21, 24
parapsychology, 174
participating consciousness, 53-54, 56
past as causal, 138-45, 146
Peacocke, A., 51, 56, 64, 65, 76-77, 79-80, 162
Peat, F.D., 53, 153, 155, 162, 164
Peirce, C.S., 124
perception, limits to, 50-51
photon, 37
Planck, M., 29
Planck's constant, 45, 46
Playfair, G.L., 168
Polanyi, M., 22-23, 28, 55
Polkinghorne, J., 64, 65
pollution, 106, 108, 109, 113
  from electromagnetic radiation, 131
Popper, K.R., 18-19, 20
positivism, 18, 55
  opposed by Kuhn's ideas, 20-21, 23
"presentational immediacy," 140-41
Pribram, K., 171-72
process thought, 3, 56-65, 80-82, 88-89, 94-95, 174
  and contemporary science, 65, 80
  and football analogy, 65-72
  see also Whitehead, A.N.
progress and evolution, 76, 85-86, 88-89, 95, 99-104
psychoanalysis, 173
purpose in cosmos *see* teleology

quanta, 32-33
quantum gravity, 46
quantum mechanical theory, 32-34
  indeterminacy *see* indeterminacy
"quantum potential" (Bohm), 153, 159
quantum wave, 34

"radial" energy, 133
Randall, J.L., 166
randomness of evolution, 75, 78, 98, 99
rationalist approach, 15, 16, 17-18
Ratzsch, D., 32

realistic approach of Einstein, 34-35, 38-39, 40
reality
  mistaking models for, 8-9, 17, 18, 25-26, 28
  and observation, 16, 17-18, 20, 31
reenchantment, 59, 62, 104, 111
regeneration, 115, 126
regulation, 115-16, 127
relatedness, importance to Whitehead, 63, 64
relativistic world, 34-36
relativity
  general theory, 34, 35
  special theory, 34
religion and science, 11, 12, 13-14
  Draper and White, 10-11
  Galileo and the Church, 8-9
  Lovelock, 113-114
  mutual dependence, 26-27
  paradigms, 13, 19-21, 24
  science in religious terms, 14-17, 24, 25-29
Russell, A., 49-50
Russell, J.B., 10
Russell, R.J., 48, 79, 164-65

saltations, 76, 85
Schrodinger, E., 39-40
science
  falsifiability, 19
  geology as an example, 29-31
  philosophy of, 17-18
  in religious terms, 14-17, 24, 25-29
  see also religion and science
scientific revolutions, 19-22
self-determination, 81, 88, 94, 143, 146, 148, 149
sense of identity, 162-63
Sheldrake, R., 114-30, 137, 144, 157, 159, 160, 166, 167, 174
Sherburne, D.W., 57, 58, 59, 141
singularities, 35, 36, 45
societies in Whiteheadian thought, 149
"soul-field," 149
spacetime continuum, 34, 35
special theory of relativity, 34

Spencer, H., 73, 86
spider webs, 127-28
sports statistics, 100-102, 128-29
Steele, J., 169
Stoeger, W.R., 74, 166
subjectivity in Whitehead's scheme, 63, 67, 69
sufficient reason, principle of, 48
syncretisms and synergisms, 156-60

$t = 0$, 45
Talbot, M., 159, 171-72
"tangential" energy, 133
Teilhard de Chardin, Pierre, 132-37
teleology
    after-the-fact argument, 84
    existence of God inferred, 42-45, 82-83, 84-85, 90
    in Gaia hypothesis, 110-11
    and natural selection, 27-28, 75, 77-78, 92-93
    non-theistic arguments, 90-93
    purpose in process thought, 68-69, 81, 88-89, 94-95
    Teilhard de Chardin, 134-37
    see also God
telos, 86-87
Templeton, J., 150
termites, 127
theology
    role, 28
    similarity to scientific method, 13-14
theory, scientific, 19
    formation of, 24-25
    paradigms, 13, 19-21
    and reality, 8-9, 17, 18, 19-22, 28
time-travel, 180-87
Tipler, F., 29, 37, 38, 42, 44, 46, 90
Torrance, T.F., 14-15, 16, 17, 22, 24-25

Toulmin, S., 27, 29
transcendence, 36, 52, 59, 126
truth, and religion, 27

ultimacy, 49, 52
uncertainty principle, 37
uniformity principle, 30
universe, history
    beginning, see Big Bang
    early years, 36, 45-46, 74, 156
    later years, see evolutionary theory

Valenza, R.J., 124
van Huyssteen, W., 13-14, 19, 21, 24
variations, adaptive see adaptive variations
vitalism, 116, 117-18, 125-26

Waddington, C.H., 119, 126
waste in evolution, 80, 87, 89, 93
Watson, L., 97, 128, 156, 162-63, 166-67
waves, in quantum theory, 34, 36-37, 78
Wheeler, J., 34, 40
White, A.D., 10-11, 53
Whitehead, A.N., 3, 8-9, 11, 56-59, 63, 64-65, 124, 137
    on causality of past, 139-41, 144-45
    concept of God, 59, 64, 68, 70-71, 80-82, 88-89, 94-95, 159-60
    see also process thought
Wigner, E., 40
wonder in scientific inquiry, 25
worldview, see field of vision, limited; paradigms

Zubiri, X., 12
Zukav, G., 53, 56

About the Author

HERB GRUNING, PH.D., teaches a variety of courses in the area of Religion and Philosophy at a number of universities in Ontario, Canada. His previous book is entitled *How in the World Does God Act?*

Printed in the United States
203224BV00004B/58-90/A